CUMULATIVE TRAUMA DISORDERS

A PRACTICAL GUIDE TO PREVENTION AND CONTROL

Wayne F. Peate, MD • Karen A. Lunda

Government Institutes
Rockville,

Government Institutes, Inc., 4 Research Place, Rockville, Maryland 20850, USA.

Copyright ©1997 by Government Institutes. All rights reserved.

01 00 99 98 97   5 4 3 2 1

No part of this work may be reproduced or transmitted in any form or by any means, electronic or mechanical, including photocopying, recording, or any information storage and retrieval system, without permission in writing from the publisher. All requests for permission to reproduce material from this work should be directed to Government Institutes, Inc., 4 Research Place, Rockville, Maryland 20850, USA.

The reader should not rely on this publication to address specific questions that apply to a particular set of facts. The authors and publisher make no representation or warranty, express or implied, as to the completeness, correctness, or utility of the information in this publication. In addition, the authors and publisher assume no liability of any kind whatsoever resulting from the use of or reliance upon the contents of this book.

Library of Congress Cataloging-in-Publication Data

Peate, Wayne F.
    Cumulative trauma disorders: a practical guide to prevention and control / by Wayne F. Peate
        p.   cm.
    Includes bibliographical references and index.
    ISBN: 0-86587-553-7
    1. Human engineering. 2. Overuse injuries--Prevention. 3. Work environment.   I. Title.
T59.7.P43   1997
620.8' 2--dc21
                            96-37429
                            CIP

Printed in the United States of America

Summary of Contents

Table of Contents v
List of Tables and Figures xi
Foreword xv
   Irving Tabershaw
Foreword xvii
   Susan J. Isernhagen
Preface xxiii
About the Authors xxv
Acknowledgments xxvii
Disclaimer xxviii

Introduction 1

Part I: Understanding CTDs and Their Causes . . . 7

   Chapter 1. What Are CTDs? 9

   Chapter 2. How Do Cumulative Trauma Disorders
      Develop? 27

   Chapter 3. What Are the Risk Factors for CTDs? 39

Part II: CTD Prevention and Control 57

Chapter 4. Fitting the Worker to the Work 59

Chapter 5. Evaluating the Demands of the Job:
Worksite Functional Job Analysis 79

Chapter 6. How to Fit Work to the Worker 117

Chapter 7. Diagnosis, Prevention, and Treatment of
Common CTDs 137

Chapter 8. Symptom Prevention 175

Part III: CTD Programs 183

Chapter 9. Proposed OSHA Ergonomic Protection
Standard 185

Chapter 10. Evaluating the Costs, Cost-Benefit, and
Cost-Effectiveness of a Program 193

Chapter 11. Choosing a Consultant 201

Chapter 12. A Sample Program 205

Glossary 215
Index 229

Table of Contents

List of Tables and Figures xi
Foreword xv
    Irving Tabershaw
Foreword xvii
    Susan J. Isernhagen
Preface xxiii
About the Authors xxv
Acknowledgments xxvii
Disclaimer xxviii

Introduction 1

Part I: Understanding CTDs and Their Causes: 7

    Chapter 1. What Are CTDs? 9

        Chapter Objectives, 9
        Case Study, 9
        Definitions, 10
        What Is Cumulative Trauma?, 11
        CTD Causes, 11
        How Common Is It?, 11
        What Has Changed?, 13
        How to Identify Risk Factors Contributing to CTDs, 13
        Why Are CTDs More Common Now?, 15
            The Written Word, 15
            Fashion Not Function, 16
            Life in the Fast Lane, 17

Combination of Factors, 18
Out of Shape, 18
The Entitlement Society, 19
Is It Real?, 19
What Does It Cost?, 20
Widespread Interest, 21
Lessons from the Past, 21
Putting It Together, 22
Endnotes, 23

Chapter 2. How Do Cumulative Trauma Disorders Develop? . 27

Chapter Objectives, 27
Case Study, 28
How Do Cumulative Trauma Disorders Develop?, 28
Function, Structure, and Response of the
 Muscle-Tendon Unit, 30
Delayed Response, 30
Symptoms and Conditions, 32
 Carpal Tunnel Syndrome, 32
 de Quervain's Syndrome, 33
 Ulnar Nerve Entrapment, 33
 Wrist Tendinitis, 33
 Epicondylitis, 34
 Rotator Cuff Tendinitis, 34
 Bicipital Tendinitis, 34
 Acromioclavicular Syndrome, 35
 Vibration Syndromes, 35
Putting It All Together, 37
Endnotes, 37

Chapter 3. What Are the Risk Factors for CTDs? 39

Chapter Objectives, 39
Case Study, 40
Biomechanics, 40
Signal Risks, 42
How Is Worker Risk Assessed?, 44
 Focus on Tasks, 44

    Review Workers' Practices, 45
    Inquire about Work Tools, 45
    Ask about Shifts, 45
    Focus on Past History, 46
    Conduct a Work Survey, 46
    Position of Body, 47
    Forceful Exertion, 47
    Time, 47
    Vibration, 48
    Handedness, 49
    Temperature, 49
    Gender, 49
    Tilted or Slippery Work Surfaces, 50
    Visual Demand, 50
    Organization of Work, 50
    New Work, 51
    Variability of Tasks, 51
    Type of Grip, 51
    Contact Stress, 52
    Improve Postures, 52
    Hobbies and Pastimes, 53
Control Opportunities, 53
Putting It Together, 54
Endnotes, 55

Part II: CTD Prevention and Control 57

Chapter 4. Fitting the Worker to the Work 59

Chapter Objectives, 59
Case Study, 60
Worker Selection Programs, 60
    Extrinsic versus Intrinsic Factors, 61
    Worker Training, 61
    Cautions on Worker Selection Programs, 62
    Preplacement or Postoffer Exams, 63
    Strength Testing, 63
    X-Rays, 64

Electrodiagnostic Studies, 64
Americans with Disabilities Act, 65
    Accommodations, 66
    Impact of the ADA, 70
    What Can One Ask about a Worker's Disability under the ADA?, 71
Putting It Together, 76
Endnotes, 77

Chapter 5. Evaluating the Demands of the Job: Worksite Functional Job Analysis 79

Chapter Objectives, 79
Case Study, 80
Worksite Functional Job Analysis, 80
Worksite Design and Ergonomic Assessments, 81
The Revised NIOSH Lifting Guide, 82
    Lifting Index, 82
    Recommended Weight Limit (RWL), 82
    Load Constant (LC), 83
    Horizontal Multiplier (HM), 83
    Vertical Multiplier (VM), 83
    Distance Multiplier (DM), 83
    Asymmetric Multiplier (AM), 83
    Coupling Multiplier (CM), 83
    Frequency Multiplier (FM), 84
    Strain Index, 84
Problem Solving, 91
    Foot Supports, 91
    Hips and Knees, 95
    Back Support, 95
    Wrists and Arms, 95
    Head, Neck, and Shoulders, 103
Completing the CTD Puzzle, 103
Assessing the Work, 104
Assessing the Worker, 105
Assessing the Worksite, 105
    Space, 106

Layout, 106
Chairs, 107
Keyboards, 107
Lighting, 108
Organization, 109
Nonphysical Issues, 109
Low Cost or No Cost Modifications, 110
The Cumulative Nature of CTDs and Back Injuries, 111
　Prevention, 111
　Functional Capacity Evaluations, 113
　Points to Consider, 114
Putting It Together, 114
Endnotes, 115

Chapter 6. How to Fit Work to the Worker 117

Chapter Objectives, 117
Case Study, 118
Work Site Design and Ergonomics, 118
Prevention Strategies for Upper Extremities, 122
　Reduce Repetition, 122
　Alter Force Required, 123
　Improve Posture, 124
　Decrease Contact Stress, 124
Prevention Strategies for Lower Extremities, 125
　Alternate Options for Standing, 125
　Avoid Kneeling/Squatting/Crouching/Stooping, 125
　Revise Sitting, 126
Evaluation of Hand Tools and Gloves, 126
　Power Grip Design, 126
　Improve Force Characteristics, 127
　Use Gloves, 128
Vibration, 128
　Localized Vibration, 128
　Whole-body Vibration, 129
Work Organization, 130
　Reduce Bending, 130
　Decrease Twisting, 130
　Revise Reaching, 130

     Lesson Lifting and Lowering, 131
     Reduce Reach, 131
     Decrease Pushing and Pulling, 131
     Minimize Carrying, 131
     Computer or Visual Display Terminals (VDT), 134
     Ergonomics Checklists, 134
Putting It Together, 134
Endnotes, 135

Chapter 7. Diagnosis, Prevention, and Treatment of Common CTDs 137

Chapter Objectives, 137
Case Study, 138
Diagnosis and Treatment Options, 138
Diagnoses and Treatment of Specific CTDs, 139
     Tension Neck Syndrome or Tension Myalgia, 139
     Cervical Strain, 140
     Epicondylitis, 141
     Tendinitis, Bursitis, and Tenosynovitis, 141
     Rotator Cuff Tendinitis or Supraspinatus Tendinitis, 142
     Carpal Tunnel Syndrome (CTS), 143
     Ulnar Tunnel Syndrome, 146
     Upper Extremity Nerve Entrapment Syndromes, 147
     Thoracic Outlet Syndrome (TOS), 148
     Bicipital Tendinitis, 149
     Acromioclavicular Syndrome, 149
     Frozen Shoulder or Adhesive Capsulitis, 149
     Pronator Syndrome, 150
     Radial Tunnel Syndrome, 150
     Posterior Interosseous Syndrome, 150
     Anterior Interosseous Syndrome, 151
     Cubital Tunnel Syndrome, 151
     de Quervain's Syndrome, 152
     Vibration Syndromes, 152
     Lumbar Sacral Syndrome, 153
Return-to-Work Issues, 166
Putting It Together, 168
Endnotes, 169

Chapter 8. Symptom Prevention 175

    Chapter Objectives, 175
    Case Study, 175
    Splints, 176
    About Computers, 176
    Supports Under Supported, 177
    Cost Effectiveness of Back Belts, 178
    CTD Exercises, 178
        To Exercise or Not to Exercise, 179
    CTD Prevention at Home, 179
    Putting It Together, 180
    Endnotes, 180

Part III: CTD Programs 183

Chapter 9. Proposed OSHA Ergonomic Protection Standard 185

    Chapter Objectives, 185
    Case Study, 186
    The Proposed Standard, 186
    General Duty Cause, 187
    What to Do if OSHA Inspects?, 188
    Putting It Together, 190
    Endnotes, 191

Chapter 10. Evaluating the Costs, Cost Benefit, and Cost Effectiveness of a Program 193

    Chapter Objectives, 193
    Case Study, 194
    What Should You Do?, 194
    Is Management Aware of the Connection Between Gender and Cumulative Trauma Disorders?, 195
    Costs, 196
    Putting It Together, 200
    Endnotes, 200

Chapter 11. Choosing a Consultant **201**

    Chapter Objectives, 201
    Case Study, 201
    A Consumer's Guide to CDT Consultants, 202
    Putting It Together, 203
    Endnote, 204

Chapter 12. A Sample Program **205**

    Chapter Objectives, 205
    Case Study, 205
    An Approach to CTDs as a Management Problem, 206
    Methods, 207
    Results, 208
    Conclusion, 210
    The Nominal Group Technique, 212
    Putting It Together, 213
    Endnote, 214

Glossary 215
Index 229

List of Tables and Figures

Figure 1.1. CTD Facts 13
Figure 1.2. Controversies Regarding Causes of CTDs 20
Figure 2.1. Checklist of CTD Factors 29
Figure 2.2. CTD Symptoms 31
Table 2.1. Common Conditions and Features of CTD 36
Figure 3.1. CTD Prevention Exercises 42
Figure 3.2. OSHA Signal Risks 43
Figure 3.3. Standard Form for Assessing Posture 52
Figure 4.1. Fitting the Worker to the Work 65
Figure 4.2. Post-Offer/Replacement Medical Evaluation 67
Figure 4.3. Physical Capacities Form 68
Figure 4.4. Key Points of the Americans with Disabilities Act 72
Figure 4.5. ADA Flow Chart 74
Figure 4.6. Sample Job Description (Essential and Marginal Funtions) 75
Figure 5.1. Revised Guide and Software Vendors 84
Figure 5.2. The feet are well supported 86
Figure 5.3. Checklist for Computer/Clerical Workstation 88
Figure 5.4. Diagram of ideal computer workstation 90
Figure 5.5. The primary problem 92
Figure 5.6. A worker behavior 93
Figure 5.7. Proper foot support 94
Figure 5.8. A slouched posture 96
Figure 5.9. Good back support along with good foot support 97
Figure 5.10. Good support and positioning of the wrists 98
Figure 5.11. The left and right wrist 99
Figure 5.12. For most people 100
Figure 5.13. Often, if the screen is too far away 101
Figure 5.14. A monitor that is too close 102

Table 5.1. Keyboard Layout 108
Table 5.2. Low Cost Approaches to Ergonomic Intervention 110
Table 5.3. Work Techniques to Prevent CTDs 113
Figure 6.1. Dimensions and Angles of Work Surfaces Form 121
Table 6.1. Control Strategies 122
Figure 6.2. Ergonomics Checklist 132
Table 7.1. Conditions That May Be Associated with CTDs 139
Figure 7.1. Carpal Tunnel Syndrome Prevention Measures 144
Table 7.2. Diagnostic, History, Physical Examination Studies and Tests for
 Carpal Tunnel Syndrome 146
Figure 7.2. Risk Factors for Nerve Entrapment 147
Figure 7.3. AHCPR Guidelines for Treating Uncomplicated Back Pain ... 153
Table 7.3. Comparison of Work-Related and Non-Work-Related Injuries .. 155
Table 7.4. Common Symptoms and Related Conditions in Low Back Pain . 159
Figure 7.4. Guidelines for Special Studies for Low Back Pain 162
Figure 7.5. Return to Work Form 167
Table 8.1. Voice Recognition Systems 177
Figure 10.1. Consultant Comparison Chart 197
Figure 10.2. CTD Cost Calculation Form 198
Table 12.1. Team Problem-Solving Areas 209
Figure 12.1. Conducting a NGT Session 213

Foreword

Cumulative trauma disorders present a challenge and an opportunity to all members of the work force and professionals. The challenges are obvious: To address disorders which often lack objective findings and clearly defined treatment guidelines.

The opportunities are clearly presented in this book: The assessment of cumulative trauma disorder prone activities and work stations in an effective manner; the development of collaborative efforts between management, healthcare providers, and workers; and the implementation of common sense approaches.

Peate and Lunda comprehensively and carefully craft for the reader an approach that deals with all affected parties involved with work injuries. You will find their text highly readable and solution oriented. They have accomplished a remarkable task. Their book is more than a list of cumulative trauma diseases and their treatments. Rather, it is a practical handbook to guide lay persons and professionals alike in their quest to not only treat, but more importantly to prevent, one of the most common groups of disorders in the workplace.

Most importantly, they address the new relationships in healthcare, those that include not only the traditional provider-patient (worker), but also the manager and employer. For in the modern workplace, treating a patient in a clinic, without understanding the dynamics and hazards of their work, represents a limited part of the injury and illness picture.

If you want to know how a worker has been injured, go to their work place, to paraphrase Bernardo Ramazzini, 18th century physician and founder of occupational medicine. Dr. Peate and physical therapist Lunda advise the same as they carefully guide the reader through the process of analyzing how workers become injured and methods by which to keep them out of harm's way.

Their passion is prevention, as will be yours once you have finished *Cumulative Trauma Disorders: Prevention and Control.*

Irving Tabershaw, M.D.
Past President of the American College of Occupational and Environmental Medicine

Foreword

It is a professional honor to address as comprehensive and fascinating a subject as cumulative trauma. It is also an honor to introduce this excellent text on the subject.

Cumulative Trauma Disorders have always been accepted in sports medicine, in music, and in the general health system. It is acknowledged that repeatedly throwing a baseball might injure the rotator cuff, that statically pinching a violin bow might exacerbate tendonitis, or that a woman with familial heredity of small wrists may develop carpal tunnel syndrome.

Therefore the existence of cumulative trauma disorders is not disputed. It is the phenomenon of increased reporting of compensable cumulative trauma disorders relating to the workplace that creates the furor in the medical and industrial worlds. The continued rise in claims have created responses in different professional sectors. This has resulted in five methods of analysis and intervention for Cumulative Trauma Disorders (CTDs).

- *The Denial Method.* A number of medical and industrial experts are taking legal stands stating that cumulative trauma disorders that are work related do not exist. Because of the relationship between compensability and CTD, experts condemn those who would be paid for reporting or treating CTDs. Because there is not absolute proof that risk factors directly cause problems, these experts can, with some logic, convince the legal system that there is not scientific evidence to exactly link stressors in the workplace with cumulative trauma reports. One strong factor in their case is always that many workers continue to be healthy and productive in the face of the same stressors.

While working primarily in the legal world to reduce costs to payers or those being sued by a government agency, the deniers have provided food for thought. They have made strides in raising the red flags regarding overreporting, but their contribution towards solutions is nil.

- ***The Medical Model.*** By analyzing the neuromusculoskeletal units involved in inflammation, nerve entrapment, scar tissue and microtears, medical specialists have been able to identify that physical phenomenon are present with cumulative trauma disorders. In many cases, however, the combinatory aspects of stresses or the exact level of damage is not easily identifiable.

Symptoms often present first with the medical signs identified later. Therapists who have traditionally worked with these diagnoses can identify trigger points, tender spots, tendon glide issues, and functional positional relationships that have a basis both in science and logic. Severe medical intervention such as surgery for carpal tunnel syndrome has also been identified as successful in some advanced cases. Device companies have proliferated splints, braces, etc. in order to "sell" the concept of medical-like intervention for prevention.

Thus, the medical world has embraced the medicalness of cumulative trauma disorders. Most practitioners feel their patients do have true conditions but these medical providers often try to control the situation with medical management only. This has provided frustration as reinjuries occur. Medical management alone has not appeared to be enough.

- *The Ergonomic and Engineering Focus.* Since cumulative trauma disorders that are work related begin at the workplace, and because the payer for work related claims ultimately is the employer, much focus has been placed on re-designing the workplace to eliminate stressors of motion or force that could create cumulative traumas. Engineers who design plants have begun to consider the human for which the equipment and worksite is designed.

    Ergonomists, who specialize in matching the work to the worker, consider human abilities and limitations to create ergonomic situations that minimize movements, forces, stresses, and non-neutral postures. This provides a safer and more comfortable worksite. The best "ergonomists" are those who work on the job. Thus the relationship between professionals and workers has fostered identification and modification of work and workplaces.

    In the zeal to quantify the workplace, measurements have been put into "formulas" which then can be multiplied, added, or mathematically entwined to provide "factors" regarding design. The mathematical model is implemented through engineering. The difficulty with only measuring forces, repetitions, pinches, grips, vibrations, and degrees of temperature is that it does not take into consideration the capacity of the worker involved in doing or absorbing the stressors.

    Implementation of ergonomics utilizes analysis of the workplace and work effects on the body. The wide variation in human workers, however, requires ergonomics and engineering to incorporate pertinent aspects of the medical/physiological model.

- *The Regulation Attack.* Whenever a problem catches the public or the payer eye, government and organizations appear ready to apply "rules and regulations." There has been a strong attempt, particularly in the United States and Canada, to reduce cumulative trauma disorders through law. By mandating employers to adhere to certain principles, the law attempts to prevent harmful conditions. This only works when the laws are logical and create positive atmospheres for reducing traumas.

A logical approach to the government's advice to industry can be found in OSHA's "Ergonomic Program Management Guidelines for Meatpacking Plants" in the United States. It gives strong advice for the evaluation of the factors that create cumulative trauma disorders. The plan incorporates employer commitment, hazard analysis, tools, engineering design, education and training, early intervention, modified work, and medical management. It provides a systems approach to analysis and intervention rather than the usual regulatory approach.

Unfortunately, other regulatory bodies and committees have tended to want to count "pounds, repetitions and minutes". Many employers, however, feel this is so potentially restrictive that industries will not be able to function. A negative aspect of the regulatory approach is that regulators become too prescriptive. A positive aspect is the fostering of a management style that involves industry, workers, and medical teams working together for evaluation and solutions.

- ***The Inclusive Approach.*** This is the approach fostered by this text!

The authors of this text have been exemplary in not believing in "easy answers" to cumulative trauma disorder questions. They clearly express belief that multiple factors in humans, in work design, and in workplace tools all create the potential for neuromusculoskeletal injury.

They have also thoughtfully included motivation, problem solving, and interactive approaches. Both workers and employers are treated with respect in this text. This underlies any solution to industrial cumulative trauma.

In addition, the authors acknowledge that cumulative trauma disorders are not exclusively work related. Personal risk factors and non-work related factors must also be considered. This human approach will provide an excellent cornerstone for a full understanding of the issues. The myriad information and references found in this text provide the reader with a logical, workable method of understanding and problem solving cumulative trauma disorders.

The bottom line is that cumulative trauma disorders will not disappear. They have always been present. They always will be present. Rather, increased understanding of the forces and human factors that create them will give us an opportunity to reduce their impact. The workers doing repetitive work will win, the employers paying the bills will win, the medical profession will broaden their knowledge from the clinic into the real world, and ergonomists will find a greater future as a result.

Appreciation is due to the primary and contributing authors for their work in bringing to us this enlightened text. Work provides a human's "right to competence." Prevention of cumulative trauma disorders is one step we professionals can take to foster safe, productive work.

Susan J. Isernhagen, PT
Isernhagen and Associates, Inc.
Author, Work Injury: Management and Prevention

Preface

Cumulative trauma disorders, or CTDs, have been described as the workplace epidemic of the late 20th century. The Occupational Safety and Health Administration (OSHA) and the National Institute for Occupational Safety and Health (NIOSH) report that CTDs are one of the most rapidly growing causes of work-related illnesses. Unfortunately, the causes of CTDs are many, and the layperson may find his or her prevention and control to be complex. Our objective is to provide the reader with the tools—in a practical, case-based format—to understand CTDs, whether at work or in the home.

CTDs are of obvious interest, not only to health care professionals who treat injured workers, but also to managers who wish to prevent these injuries. To determine if educating managers in sound ergonomic principles in the workplace could have an effect on the number of CTD cases, we developed and instituted a CTD prevention training program, as described in Chapter 12. We then wrote this book because we could not find a reliable reference to use in our training program. We are confident you will find our efforts to be helpful as you seek to address the problem of CTDs wherever they occur.

Wayne F. Peate, MD, MPH
Karen Lunda, MS, PT

About the Authors

Wayne F. Peate, M.D., M.P.H. is an assistant professor of clinical family and community medicine at the Arizona Prevention Center, University of Arizona, and serves as the director of the Environmental/Occupational Health Unit. He has received the University of Arizona, College of Medicine, department award for excellence in teaching. Dr. Peate received his M.P.H. at the Harvard School of Public Health, his M.D. at Dartmouth Medical School and an A.B. at Stanford University. He is board certified in preventive medicine (occupational medicine) and medical management, is a fellow of the American Academy of Disability Evaluating Physicians and is a Federal Aviation Administration Medical Examiner. A member of the University of Arizona, College of Medicine faculty since 1984, Dr. Peate was Medical Director for Halba Refugee Camp, Somalia from 1981 to 1982, and has served as a past president of the Harvard School of Public Health Alumni Association and as a consultant for Fortune 500 companies. He has published articles on burns, the health hazards of travel, medical management, cumulative trauma disorders, and musculoskeletal diseases.

Karen Lunda, M.S., P.T., who contributed to Chapters 1 and 5, has over thirteen years of experience in physical therapy. The past eight years have been in industrial rehabilitation. She holds a B.S. in physical therapy and an M.S. in exercise physiology. She was instrumental in creating and developing the Carondelet Injury Prevention and Management Program at Carondelet Health Care in Tucson, Arizona, where she is a physical therapist. She has significant experience in the areas of work hardening, work conditioning, functional capacity evaluation, job site analysis, ergonomics, and Americans with Disabilities Act (ADA) consultation. She has given presentations on injury prevention and management at the local, state, and national levels, and has worked closely with the Commission on Accreditation of Rehabilitation Facilities (CARF) as a surveyor, an assistant to the CARF staff in the training of new surveyors, and as a member of two national advisory committees. She is published in the areas of ADA and ergonomics.

Acknowledgments

The authors especially thank Cindy Levack for her invaluable contributions in the preparation of this manuscript, along with Naomi Neff, Trisha Stanley, Patsy Moreno and Denise Gingerich for their help.

Thanks also go to Alex Padro, Dorothy Foster Rea, and Joel Moorhead, M.D., M.P.H. for assistance in editing.

Special gratitude to Lynn M. Struthers Peate, who survived the "3 Bs" this year: new book, new baby, and new business, and my daughters, who lost precious hours with me at home because of "Daddy's working."

Disclaimer

This book is designed to offer items of general interest to the reader. It is not addressed to individual problems and is not intended to offer medical or legal advice. A local health care practitioner should be consulted for advice on medical conditions, and an attorney for legal and regulatory issues.

The authors specifically disclaim any liability, loss, or risk, personal or indirectly of the use and application of any of the contents of this book.

CUMULATIVE TRAUMA DISORDERS

A PRACTICAL GUIDE TO PREVENTION AND CONTROL

Introduction

"Discovery consists of looking at the same thing as everyone else and thinking something different."
—Szent Gregory, Nobel Prize winner.

Objectives

After reading this book, the reader will:

- Understand current recommendations for the prevention and control of cumulative trauma disorders (CTDs).
- Be able to describe cost-effective approaches for the prevention of CTDs at home and in the workplace.
- Understand what is meant by ergonomics as it relates to CTD prevention and control.
- Understand the model of CTD self-responsibility and the worker as a partner.
- Recognize that dealing with CTD symptoms is less effective than addressing the conditions that lead to CTDs.
- Be aware of the fundamental principles of job analysis, including applying the NIOSH strain and lifting index to simple tasks to prevent and control CTDs.

- Know the limitations of knowledge related to the causes of CTDs, and that the field of CTD prevention and control is always changing with new information.
- Understand the controversies related to the use of splints, back supports, and CTD prevention exercises.

What This Book Is About

Our purpose is to help the reader prevent the preventable—cumulative trauma disorders (CTDs) in the workplace and in the home. Our plan is to provide the reader with a readable, practical, and up-to-date road map to prevent and control CTDs.

Scope of the CTD Problem

In 1995, Joseph A. Dear, Assistant Secretary of Labor for Occupational Safety and Health, issued the following statement: "Work-related musculoskeletal disorders affect more than 700,000 workers every year and account for an incredible one out of three dollars spent on workers' compensation."[1] The Bureau of Labor Statistics (BLS) defines "repetitive trauma"—or CTD—as a condition due to repeated motion, pressure, or vibration. CTDs have been variously described as repetitive motion injury, repetitive strain, and even "Kangaroo Paw" during Australia's CTD epidemic in the 1980's.[2]

In 1993, CTDs accounted for nearly two-thirds of total occupation-related illnesses in the United States, 5 percent of all injuries and illnesses reported by private industry, these injuries surpassed the 1992 total by 7 percent, and increased from 100 cases per 100,000 in 1987 to 383 cases per 100,000 workers.

The industries with the highest numbers of CTDs in 1993 included motor vehicle and equipment manufacturers with 42,600 cases (40,600 in 1992), meat packing with 38,300 (36,500 in

1992), and aircraft and parts manufacturing with 9,500 cases accounted (8,600 in 1992), according to BLS. The BLS surveys are based on data compiled from the OSHA 200 injury and illness logs maintained by industries with one or more employees.[3,4]

Rapid Response

It is critical to get involved early and proactively in CTD prevention and control. The longer workers stay off work due to illnesses such as CTDs, the less likely they are to ever return.[5]

Why This Book?

Employers often lack the knowledge and do not know who to call or when to seek outside help to reduce the risk of CTDs in the workplace.[6] Some may be unaware of the expense of various options that are equally effective. A recent study regarding the treatment cost of back injuries, considered a form of CTD, found that primary care providers, in contrast to others, offer similar outcomes at less cost. The average cost of care given by primary care providers was $365 to $540, by orthopedists $809, and by chiropractors $505 to $808.[7] Others attempt to prevent or control CTDs with various modalities, such as back belts, wrist supports, stress reduction classes, exercises, and strength testing programs whose efficacy may be controversial.[8,9]

Who Wrote This Book?

An occupational medicine, physician and a physical therapist with special training in CTDs, job analysis, ergonomics and are the authors of this book. Our intended audience is broad, including human resources personnel, safety and health professionals, ergonomists, managers, and workers. We will attempt to address the needs of our entire audience, because we recognize that some readers will have prior experience with CTDs, while many will not.

A Road Map

To aid the reader, each chapter will contain:

- Tables and boxed sections that explain the high points
- Cases with real life examples.
- Concise summaries of supporting research from leading journals.
- A "Putting It Together" section—tips for practical use—at the end of each chapter.
- Guidelines, checklists, and forms.
- Pictures and diagrams of actual work stations to aid understanding of CTDs.
- Chapter objectives—what the reader can accomplish by following the recommendations and gaining the knowledge offered in the book.
- Relevant quotes to open each chapter.
- A bias: we believe in self-responsibility and the worker as a partner. All of the ergonomic changes and CTD prevention programs available are not effective if they are never used, and without worker cooperation any interventions will be severely limited. Our experience is that workers frequently have the best and often the most cost-effective suggestions for CTDs prevention and control. Research has shown that jobs associated with high demands and little control lead to physical complaints,[10] so allow the best ideas on how to control CTDs to flow from workers.

Endnotes

1. *ACOEM Report*. Marianne Dreger (ed). American College of Occupational and Environmental Medicine, July 1995.
2. Woolard, T.J. Occupational overuse syndromes—The Australian experience. Repetitive strain disorders. *Seminars in Occupational Medicine* 2:7-10, 1987.
3. National Safety Council. *Accident Facts*. 1993 edition. Itasca, IL: National Safety Council, 1993.
4. National Safety Council. *Accident Facts*. 1994 edition. Itasca, IL: National Safety Council, 1994.
5. Waddell, G. A new clinical model for the treatment of low back pain: 1987 Volvo Award in Clinical Science. *Spine* 12:632, 1987.
6. Maizlish, N., Rudolph, L., Devin, K., and M. SanKaranarayan. Surveillance and prevention of work-related carpal tunnel syndrome: an application of the Sentinel Events Notification System for Occupational Risks. *American Journal of Industrial Medicine* 27:715-729, 1995.
7. Carey, T.S., Garrett, J., Jackman, A., McLaughlin, C., Fryer, J., and D.R. Smucker. The outcomes and costs for acute low back pain among patients seen by primary care practitioners, chiropractors and orthopedic surgeons. *New England Journal of Medicine*, October 5, 1995.
8. Alexander, O. et al. The effectiveness of back belts on occupational back injuries and worker perception. *Professional Safety* 9:22-26, 1995.
9. Dueker, J.A., Richie, S.M., Knox, T.J., and S.J. Rose. Isokinetic trunk testing and employment. *Journal of Occupational Medicine*.
10. Karasek, R. *Healthy Work: Stress Productivity and the Reconstruction of Working Life*. New York: Basic Books, 1990.

Part I
Understanding CTDs and Their Causes

"Our only limitation to our achievement of tomorrow is our doubts of today."
—*Franklin Delano Roosevelt*

Part I can be summarized as understanding the seven Ps: person, place, posture, practices, position, personality, and physical.

Chapter 1: What Are CTDs? 9

Chapter 2: How Do Cumulative Trauma Disorders Develop? 27

Chapter 3: What Are the Risk Factors for CTDs? 39

1

What Are CTDs?

"It is just as important to know what sort of person has a disease as what sort of disease the person has."
—*Hippocrates*

Chapter Objectives

At the completion of the chapter, the reader will:

- Understand what CTDs are.
- Realize how common CTDs have become.
- Understand the changes in the workplace and workforce that have led to the increase in CTDs.

Case Study

Get Help Before It Is Too Late

Mr. B is a 29-year-old window and door installer. Two days before a newly remodeled high school was set to open, an inspector noted no fire doors were ever installed. Mr. B was given the job. In an eight-hour day, he hauled fifteen 110-pound doors up two flights of stairs, using his left hand to carry the weight and his right hand to steady the doors. Within seven hours after completing the installation, he complained of left hand numbness with weakness. A subsequent medical workup revealed he had torn the triangular fibrocartilage complex at the wrist and stretched the ulnar nerve at the elbow. After two surgeries he has not improved.

Definitions

A cumulative trauma disorder is defined in the United Kingdom as a recurrent or persistent musculoskeletal pain without an immediate traumatic cause occurring within the prior 6 weeks.[1] The proposed OSHA ergonomic standard suggests that "work-related musculoskeletal disorders"—which include CTDs—are caused or aggravated by exposure to workplace risk factors, including signs or persistent symptoms of at least 7 days, interference with work, or clinically diagnosed work-related musculoskeletal disorders.[2]

The Bureau of Labor Statistics defines CTD as an illness (like asbestosis) rather than an injury (like low back pain). Others define cumulative trauma disorders as chronic injuries to the soft tissues caused by repetitious exertions.[3] CTDs are also known as repetitive strain injuries, repetitive motion injuries, wear and tear injuries, or overuse injuries. The National Institute of Occupational Safety and Health asserts that trauma is induced when "job demands... repeatedly exceed the biomechanical capacity of the worker. Sources of trauma are workplace tasks or non-work-related activities that cause repetitive and continuous strain and contribute to the onset of injuries affecting the musculoskeletal system."[4]

What Is Cumulative Trauma?

A CTD usually results from the cumulative stress created by a combination of risk factors. Cumulative trauma is not a one-time event. It is not falling off a ladder and hurting a knee, lifting a single box and sustaining a back injury, or typing one letter and getting carpal tunnel syndrome. It is the accumulation of trauma to a body part over a period of time. Trauma in this sense is not a violent force, but minor stressors that when repeated and combined can accumulate and lead to symptoms.

This cumulative effect can affect any part of the body that moves. CTDs can develop in the thumb, elbow, ankle, or other joints. Even certain back injuries are recognized as cumulative in nature. The most publicized CTD is carpal tunnel syndrome (CTS), which involves compression of the median nerve at the wrist.

CTD Causes

Workplace factors that lead to CTDs include improperly adjusted equipment, long work shifts without breaks, forceful repetitive work,[5] lack of task variability, machine-paced work, inadequate rest-work cycles, poor physical condition, awkward position, excessive force, twisting movements, poor body mechanics, posture, position,[6] localized and whole body vibration, cold environments,[7] carrying, lifting, pushing, pulling, poor lighting, various psychosocial factors,[8] and gender.[9]

How Common Is It?

Cumulative trauma disorders have been described as the work-related epidemic of the 1990s. Disorders associated with repetitive trauma were the most common occupational illnesses reported by the Bureau of Labor Statistics (BLS) in 1992, with 281,800 cases, representing 62 percent of all new occupational illnesses reported in the United States and 4 percent of the occupational injury and illness total—a decided increase over previous years.[10] In contrast, in 1981 CTDs amounted to only 18 percent of reported new occupational illnesses, about 22,700 cases.

Comparative studies with insurance company data suggest the condition is underreported.[11] Underreporting may occur for several reasons, including lack of knowledge, controversy regarding causation, and workers obtaining private treatment outside of the workers' compensation system.

In 1993 the CTD incidence rate reported by the Bureau of Labor Statistics was 8.5 for every 100 full-time equivalent workers. This rate is actually a decrease from the 1992 rate of 8.9. The Occupational Safety and Health Administration (OSHA) attributes this decline to its enforcement actions as well as to ergonomic programs implemented in industries such as meat packing and automotive manufacturing.[12]

The rise in CTDs over the prior decade was dramatic. CTDs accounted for 18 percent of reported occupational injuries in the United States in 1982 and 52 percent in 1989.[13] One investigator reports an incidence rate of industrial cumulative trauma disorders of 6.2 to 129.6 cases per 200,000 work hours, depending on specific job duties, and notes that CTD cases are probably underreported on OSHA logs.[14] Figure 1.1 contains statistical facts about CTDs.

CTDs:

- Represent half of all occupational illnesses.[15] Note that the Bureau of Labor Statistics separates cases as either illnesses (CTDs) or injuries (low back pain).
- Have increased 1,300 percent over the past 12 years.[16]
- Represent one third of all workers' compensation expenses.[17]
- Are due to placing, grasping or moving objects in 31 percent of cases, and to typing or key entry in 12 percent.[18]

Figure 1.1 CTD Facts

What Has Changed?

Cumulative trauma disorders in the workplace are not new. The condition was noted in telegraph operators and others in the past century. However, they are occurring with increasing frequency as the workplace has become more automated with repetitive duties, constrained amplitude of motion (such as with cashier, small parts assembly, and keyboard work), less frequent task rotation, and performing the same function 4,000 to 5,000 times per day.[19] All of these factors place repeated demands on the same part of the body, and more workers are employed in occupations at risk of CTDs.[20] Computerization, like the assembly line before it, has aided efficiency at the expense of the worker or operator.

How to Identify Risk Factors Contributing to CTDs

The type of work, method, and place of work may all have a bearing on whether CTDs develop. For example, why can some workers type all day at a keyboard for years and never develop a problem and others seem to develop symptoms of carpal tunnel syndrome after one or two years of only intermittent typing?

To understand the above, consider why some people develop heart disease and others do not, or why one person is cavity free and another has a mouth full of dental decay. The answer to this question can be found in risk factors. Amount of exercise, how stress is handled, the kind and amount of food eaten, our genetic code, and smoking are all variables that have an effect on cardiac status. The percent that each plays will vary from individual to individual. Sometimes one factor may play a larger role, but often it is the combination of several factors that leads to a higher level of risk. The contribution of each factor is within individual control except for one—heredity. Changing a person's genetic makeup and tendencies is not possible. (However, keep posted on advances in genetic engineering and the Human Genome Project to be completed in 2005.) We can have an impact on how much and what type of exercise we do, how much stress we have and the way we handle it, the amount and type of food we eat, and how much we smoke tobacco.

It is not easy to control these risk factors—most are ingrained in personal habits that have developed over years. These habits can be difficult to break or even change, such as the exercise program that falters in a week or two, the New Year's diet that stops by the end of January, or failed attempts to quit smoking or drinking. Most people could identify with one of the above situations.

Risk factors that contribute to CTDs in the workplace follow a similar scenario. Like heart disease, there are many factors that contribute to a CTD. Some may involve the way workers perform their work or the design of the work site in which the work is being performed. Usually changes can be made to reduce or eliminate some of the risk factors, although similar to the hereditary component of heart disease, there may be a risk factor that is difficult to change.

Why Are CTDs More Common Now?

Changes in the workplace have contributed to the rise in CTDs. Work environments of particular risk include:

- When constrained amplitude of motion with repetitive movement occurs (cashier, keyboard, assembly line).

- When the above are combined with preexisting factors (such as arthritis and decreased muscle mass more common in an older work force).

- When multiple movements are combined, such as twisting and lifting at the same time.

- When women use work station designs with average U.S. Air Force white male anthropometric data from the 1950s.[21]

- When women are increasingly employed in nontraditional jobs, such as firefighting, where equipment and tools may be of improper size. One female candidate for firefighting failed the qualifying test because the helmet provided was too large and kept slipping forward and blocking her vision.

The Written Word

Consider typists (old term for word processor persons) who 20 to 30 years ago varied their routine several times an hour, first inserting paper (rotating the elbow and grasping with the hand) by rolling it into the typewriter, then typing (small movements of the fingers and hands), then pressing the return lever (internal rotation of the shoulder and flexion of the wrist)—all involving varying movements with different body groups that were then allowed to rest.

Now examine the modern use of computers. Where once workers varied their daily activities a great deal during the course of a day there is now less variation. As work has become more automated, a computer user or an assembly line worker may move a limited set of muscles repeatedly. Clerical workers may sit at a desk for 3 to 4 hours at a time, never removing their hands from the keyboard, and completing 10,000 keystrokes an hour. Paper is already loaded in a printer that prints with the touch of a few keystrokes. The return is done automatically at the end of a sentence or by another keystroke.

There is little opportunity to relax a busy group of repetitively working muscles without a recovery or relaxation period, a situation which is aggravated by the modern trend to expect more work from fewer workers, such as in downsized operations.

A report that examined the distribution of CTD by job classification found that clerical office employees were in the top ten occupations affected with CTDs, with these workers accounting for 7.25 percent of CTD claims and 8.89 percent of CTD costs.[22]

Another often ignored group of muscles are those of the eye. In one study, subjects watching a visual display terminal reported greater fatigue than those monitoring an auditory display.[23]

Fashion Not Function

The workplace has traditionally been designed more for appearance and fashion than function—and usually for men, not women. Chairs and other parts of the work station were usually developed for a 5 foot 10 inch tall male with a work height of 29 to 31 inches, which may be inappropriate for the 43 percent of the U.S.

work force that is female.[24] The chair a worker sits in is usually inherited from a past worker, no matter that one is at the 5th percentile for height and the other was at the 95th percentile.

Not only is height involved, so is length of knee to hip, arm and trunk length, and size of tools and grips—all of which complicates fitting the worker to the work station. For example, work station design references recommended an overhead reach of 77 inches for males, whereas 73 inches is advised for 95 percent of females.[25]

CTDs also occur in nonclerical operations. A recent study found that in a group of 308 electricians over the course of a year:

- 82 percent reported musculoskeletal complaints,
- 35 percent required medical assistance and job status changes for lower back discomfort,
- 25 percent complained of neck symptoms.[26]

Life In The Fast Lane

Other factors involved in CTDs may include psychosocial stresses on the modern family. Two parents working outside the home, limited periods of relaxation (weekends consumed with catching up on family chores and tasks), concerns about possible layoffs, a poor job performance appraisal, and low job satisfaction, have all been implicated as causes of CTDs, as well as smoking, caffeine, alcohol, inadequate sleep and substance abuse.[27,28,29]

Combination of Factors

Organizational factors may also play an important role. Until recently, CTDs have been attributed largely to physical conditions. However, a recent study at a large U.S. newspaper newsroom suggests that a combination of physical and psychological factors, especially workers' relationships with their supervisors, contribute to CTDs, and that revising work stations may not be sufficient if interpersonal conflicts are an issue.[30]

The Faucett study estimated the frequency of CTD symptoms and the contribution of work posture, job characteristics, and interpersonal relationships to the severity of these symptoms. It demonstrated the combined effect of both physical and psychosocial factors on workers and concluded that work stress may worsen the effects of a poor work station and that workers with poor relationships with their supervisors had more severe CTD symptoms, even though they had ergonomically approved work stations. This study should give pause to employers who consider ergonomics or job stress the only causes of CTDs. Organizational factors such as employees' relationships with supervisors may need to be addressed first.

Out of Shape

Even general fitness, or lack of it, has been implicated in CTDs. Firefighters who sustained low back injury, but who were more fit before the injury recovered faster than those who were less fit.[31] Unfortunately, one third of the workforce in the United States is overweight, 60 percent rarely exercise and 15 percent do not exercise at all.[32] Aging also contributes to CTDs with the development of bone spurs in joints, loss of bone density, and decreased muscle mass.

The Entitlement Society

Many modern industrialized countries have made progress in protecting the rights and security of the disabled through various entitlements, such as the Social Security Act. Some attribute the rise in CTD claims to an "entitlement attitude" among modern workers. In the postdepression era, many employees felt fortunate to have a job, the alternative being deprivation. Today's workers know greater economic security and may feel less inhibited in demanding workplace changes and compensation for any perceived physical or psychological damage they might have incurred while working. Workers are increasingly aware of potential physical stressors in the workplace and are more likely to voice their concerns about any effects on their health, such as CTDs, than their predecessors did.

Is It Real?

Certainly many individuals have legitimate work-related CTDs, and employee self-report of CTD symptoms has generally been found to be accurate,[33,34] in spite of concerns about the unreliability of workers and symptom magnification of workers' compensation patients.[35-37] Increased media, government, and medical awareness have also been cited as a cause for increased reporting of CTDs.[38]

Figure 1.2 summarizes the controversies surrounding CTD causation.

- Psychological: Patients suffering from carpal tunnel syndrome (CTS) have reported a higher degree of distress and less "lifestyle organization," such as lack of efficiency and timeliness.[39] A key question—as yet undetermined—is to what degree the distress was due to having CTD or whether it preceded the symptoms.
- Media, government, and medical awareness.[40]
- Drugs: Legal drug use, to a varying degree, appears to affect the prevalence of carpal tunnel syndrome caused by cumulative trauma.[41]
- Symptom exaggeration.[42]
- Litigation: Retaining legal assistance has been associated with reduced distress in low back pain, perhaps as a coping mechanism.[43]
- Habit: Some workers press excessively while others perform the same task gracefully—and less traumatically.

Figure 1.2 Controversies Regarding Causes of CTDs

What Does It Cost?

In addition to pain and impaired physical status, significant costs are involved. A recent study in Wisconsin analyzed the cost of treating work-related carpal tunnel syndrome, a condition that may be caused by cumulative trauma. Benefits and medical expenses for the average case were $29,000. For the entire state, the yearly cost for surgical cases was $51 million. This latter figure does not include lost productivity costs, which could bring the figure to $100,000 per case for nonsurgical CTS cases.[44]

Widespread Interest

The potential for prevention of occupational CTDs makes these disorders of special interest to management as well as to the many other professionals who manage and treat injured workers. Corporate and individual management of CTDs in an occupational setting presents complex social and medical issues, including physical conditions, such as preexisting medical conditions, psychological factors, such as relationships with supervisors) and organizational issues, such as increased productivity.[45]

Lessons from the Past

Cumulative trauma disorders are not unique to the 20th century and automated working conditions. In 1713, Ramazzini noted: "There are many persons who support themselves and their families solely by writing. Incessant driving of the pen over paper causes intense fatigue of the hand and the whole arm because of the continuous and almost tonic strain on the muscles and tendons, which in course of time results in failure of power in the right hand."[46] By the 1880s, "telegraphists' cramp" was a common CTD in telegraph workers—caused by the rapid movements of a telegraph key to transmit Morse code. In 1905, Henry Ford introduced the assembly line to improve production, but it also increased the amount of physical stress on a limited group of muscles in workers. In the textile factories of England, a report in 1919 noted that cotton twisters reported pain, cramping, and weakness in affected muscles from joining ends of threads by twisting them between the index finger and the thumb. The condition was called "cotton-twisters' cramp."

Settings where workers are most at risk in the contemporary workplace are meat packing houses, pottery manufacturers, mailing or addressing companies, knit goods and hosiery manufacturing, electric power or transmission equipment manufacturing, boot or shoe manufacturing, aircraft engine manufacturing, clothing manufacturing, clerical offices, and meat or poultry dealers or stores.[47] The highest number of cumulative trauma disorders occur within manufacturing industries,[48] and the risk for women is greater than for men.[49,50]

Putting It Together

- CTDs are common, preventable, controversial, and not always distinguishable from other conditions.

- Changes in the workplace have contributed to the rise in CTDs. Work environments of particular risk include constrained amplitude of motion in repetitive movement environments, such as keyboard or assembly line work, in combination with other factors, such as arthritis, and when multiple movements are involved, such as twisting and lifting.

- Work settings, whose design was initially based on white males in the U.S. Air Force in the 1950s, may be inappropriate for women and other ethnic groups. Adjustable work areas or redesign may be indicated.

Endnotes

1. Guidotti, T. Occupational repetitive strain. *American Journal of Family Practice* 45:585-592, 1992.
2. CFR1910. OSHA Proposed Ergonomics Standard, 1995.
3. Zimmerman, N.B., Zimmerman, S.I., and G.L. Clark. Neuropathy in the workplace *Hand Clinics.* 8:255-262, 1992.
4. Association of Schools of Public Health under a cooperative agreement with the National Institute for Occupational Safety and Health. Proposed strategies for the prevention of leading work-related diseases and injuries. 1986, p 19.
5. Isernhagen, S.F. *Cumulative Trauma Prevention.* Duluth, MN: Isernhagen Work Systems 1989.
6. See note 1.
7. Pyykko, I. Clinical aspects of the hand-arm vibration syndrome: A review. *Scandinavian Journal of Work Environment and Health* 12:439-447, 1986.
8. Bigos, S.J., Battie, M.C., Spengler, D.M., Fisher, L.D., Fordyce, W.E., Hansson, T.H., Nachemson, A.L., and M.D. Wortley. A prospective study of work perceptions and psychosocial factors affecting the report of back injury. *Spine* Volume 16, June 1991.
9. Ashbury, F.D. Occupational repetitive strain injuries and gender in Ontario, 1986-1991. *Journal of Occupational and Environmental Medicine* 37:479-485, 1992.
10. National Safety Council. *Accident Facts.* 1994 Edition. Itasca, IL: National Safety Council, 1994.
11. Brogmus, G., and R. Marko. The proportion of cumulative trauma disorder of the upper extremities in U.S. industry. *In*: Proceedings of the Human factors society 36th annual meeting. October 12-16, 1992, Atlanta, Georgia. *Human Factors and Ergonomics Society.* 2:997-1001, 1992.
12. ACOEM, Marianne Meyer (ed). Report (American College of Occupational and Environmental Medicine), February 1995.
13. See note 10.
14. See note 3.
15. See note 11.
16. See note 13.
17. See note 12.
18. See note 13.
19. Margolis, W., and J.F. Kraus. The prevalence of carpal tunnel syndrome in female supermarket checkers. *Journal of Occupational Medicine.* Volume 29, December 1987.
20. Kumar, S., and A. Mital. Preface: Telecommunications. *Ergonomics* 37 (10):1,589-1,590, 1994.

21. Salvendy, G. (ed). *Handbook of Human Factors.* New York: John Wiley, 1987.
22. See note 11.
23. Galinsky, T.L., Rosa, R.R., Warm, J.S., and W.M. Dember. Psychophysical determinants of stress in sustained attention. *Human Factors.* Volume 35, December 1993.
24. Chaffin, D.B., and G. Anderson. *Occupational Biomechanics.* New York: John Wiley, 1984.
25. See note 24.
26. Hunting, K.L., Welch, L.S., Cuccherini, B.A., and L.A. Seiger. Musculoskeletal symptoms among electricians. *American Journal of Industrial Medicine*, February 1994.
27. Kelsey, J.L., Githens, P.B., O'Connor, T., Weil, V., Calogro, J., Holford, T.,White, A., Walter, S., Ostfeld, A., and W.O. Southwick. Acute prolapsed lumbar intervertebral disc. An epidemiologic study with special reference to driving automobiles and cigarette smoking. *Spine*, Volume 9, September 1984.
28. Kelsey, J.L., and M.C. Hochberg. Epidemiology of chronic musculoskeletal disorders. *Annual Review of Public Health* 9:379-401, 1988.
29. Nathan, P., Keniston, R.C., Lockwood, R.S., and K.D. Meadows. Tobacco, caffeine, alcohol and carpal tunnel syndrome in American industry. *Journal of Occupational and Environmental Medicine* 38(3) 284-298, March 1996.
30. Faucett, J., and D. Rempel. VTD related musculoskeletal symptoms: Interactions between work posture and psychosocial work factors. *American Journal of Industrial Medicine*: 597-612, 1994.
31. Cady, L.S., Bischoff, D.P., and E.R. O'Connell. Strength and fitness and subsequent back injury in firefighters. *Journal of Occupational Medicine* 21:269-272, 1979.
32. U.S. Department of Health and Human Services. *Healthy People 2000: National Health Promotion and Disease Prevention Objectives.* DHHS Publication No (PHS) 91-50213. Washington, D.C.: U.S. Government Printing Office, 1991.
33. Katz, J.N., Chang, L.C., Sangha, O., Fossel, A.H., and D.W. Bates. Can comorbidity be measured by questionnaire rather than medical record review? *Medical Care.* In press.
34. Mason, J.H., Anderson, J.J., Meenan, R.F., Haralson, K.M., Lewis-Stevens, D., and J.L. Laine. The rapid assessment of disease activity in rheumatology (RADAR) questionnaire. *Arthritis and Rheumatism* 35: 156-162, 1992.
35. Nathan, P.A., Keniston, R.C., and K.D. Meadows. Carpal tunnel syndrome in the work place. *Hippocrates' Lantern* 2:1-4, 1993.

36. Hadler, N.M. Cumulative trauma disorders: an iatrogenic concept. *Journal of Occupational Medicine* 32:38-41, 1990.
37. Higgs, P.E., Edwards, D., Martin, D.S., and P.M. Weeks. Carpal tunnel surgery outcomes in workers: Effect of workers' compensation status. *Journal of Hand Surgery—American*, Volume 20A, May 1995.
38. Brogmus, M.S., Sorock, G.S., and B.S. Webster. Recent trends in work-related cumulative trauma disorders of the upper extremities in the United States: An evaluation of possible reasons. *Journal of Occupational Environmental Medicine* 38:401-411, 1996.
39. Vogelsang, L.M., William R.D., and K. Lawler. Lifestyle correlates of carpal tunnel syndrome. *Journal of Occupational Rehabilitation* 4:141-152, 1994.
40. See note 38.
41. See note 29.
42. See note 36.
43. Tait, R.C., Chunall, J.T., and W.D. Richardson. Litigation and employment status: effects on patients with chronic pain. *Pain* 43:37-46,1990.
44. Hanrahan, L.P., Higgins, D., and H. Anderson. Wisconsin occupational carpal tunnel syndrome surveillance: the incidence of surgically treated cases. *Wisconsin Medical Journal* 92:685-689, 1993.
45. See note 11.
46. Dalton, S., and B.L. Hazelman. Repeated movements and repeated trauma. In: *Hunter's Diseases of Occupations*, 6th edition. Boston: Little, Brown, 1978: p 620-621.
47. See note 11.
48. National Safety Council, *Accident Facts*. 1993 edition. Itasca, Illinois: National Safety Council, 1993.
49. See note 9.
50. See note 38.

2

How Do Cumulative Trauma Disorders Develop?

"What a piece of work is a man, how noble in reason, how infinite in faculties, in form and moving how express and admirable."
—Hamlet, *William Shakespeare*

Chapter Objectives

After completing this chapter, the reader will:
- Be able to explain how cumulative trauma disorders develop.
- Be able to describe the causes of cumulative trauma disorders.
- Identify the types of activities that lead to cumulative trauma disorders.
- Know the basic anatomy, structure, function, and symptom responses of various parts of the body affected by cumulative trauma disorders.

- Know the risk factors for cumulative trauma disorders.
  - *Physiologic*: arthritis, congenital factors (such as wrist size), diabetes, gout, thyroid disorders, and pregnancy.
  - *Work related*: work tasks and conditions.
  - *Behavioral*: attitudes toward work and relationships with supervisors.

Case Study

Vibration

A 25-year-old general laborer complained of a five-month history of tingling in both hands, especially after a long shift using a power chipper. Symptoms were not present before use of this tool. Now he has difficulty performing tasks such as holding objects and counting change.

How Do Cumulative Trauma Disorders Develop?

Cumulative trauma disorders are not like the common cold. They cannot be caught from a co-worker. In fact, two workers doing exactly the same job may find one develops a CTD and the other does not. How then do cumulative trauma disorders develop?

Clearly, certain activities contribute to cumulative trauma disorders, such as improper body mechanics, awkward position, excessive force, repetitive movement, torquing maneuvers, increased work load, time pressure, and machine pacing,[1,2] but to what degree they contribute is often not clear. It is even more difficult to assign causation.

There are concerns that nonphysical issues may be contributing to CTDs. Examples include management issues, the workers' relationships with supervisors, mental health issues, and high demands and lack of control in the workplace. Factors that contribute to cumulative trauma disorders are outlined in the following figure. Because CTDs may be caused by more than one factor, it is critical to assess the entire work and home environment. This topic will be considered further in Chapter 3.

Figure 2.1 provides a checklist to focus your scan of possible cumulative trauma disorder factors.

ACTIVITY	POSTURE/ POSITION	TIME/ DURATION	FREQUENCY/ REPETITION	FORCE
1. sit				
2. stand				
3. walk				
4. squat				
5. crouch				
6. crawl				
7. hop				
8. reach				
9. twist				
10. grip/pinch/ power				
11. push/pull				
12. lift				
13. carry				
14. kneel				
15. climb				
16. reach above shoulder				
17. bending neck				
18. bending back				
19. keyboard use				
20. vibrating tool				
21. foot control/ pedal				
22. temperature/ humidity				
23. dust/fumes				

Figure 2.1 Checklist of CTD Factors

Function, Structure, and Response of the Muscle-Tendon Unit

A key factor in understanding cumulative trauma disorders is an awareness of the function, response, and structure of the muscle-tendon unit, which includes tendons, tendon sheaths, and entheses (the tendon-bone junctions). Each component has a distinct function, structure, and response to cumulative trauma disorders. These tendons are similar to pulleys and move back and forth in a sheath that lubricates movement. What does this pulley action have to do with cumulative trauma disorders? Many people experience muscle or joint soreness after a bout of physical activity, whether it be from playing softball after a long absence—the weekend warrior—or from an activity such as planting a lawn that requires muscles unused for some time. The causes of this discomfort are many. As fatigue increases, blood flow increases and lactic acid, a byproduct of active muscles, builds up in overused, under-conditioned muscles. The result is microtrauma to the muscle-tendon unit. These pulleys, after multiple uses, also undergo physiological changes. These include edema or swelling, and even small hemorrhages or bleeding.

Delayed Response

Physiological changes in the pulleys lead to symptoms that often are not noticeable until days later. Cumulative trauma disorders follow the same pathway. Multiple repetitive microtrauma, in contrast to obvious acute injuries such as fractures, also leads to symptoms, which may not be as pronounced or occur as quickly as in the weekend softball player, but whose effects can be more serious—even resulting in surgery and disability. In addition, muscles move joints that move while under a load or weight. Degenerative joint disease can develop. This may lead to changes such as cartilage loss or an overgrowth of bone, called a bone spur. Some consider degenerative joint disease a derivative of cumulative trauma disorders.

The symptoms of CTDs are outlined in Figure 2.2.

- Pain and discomfort in the extremities or spine.
- Numbness, tingling, a pins and needles sensation.
- Weakness.
- Decreased range of motion.
- Decreased work capacity.
- Clumsiness.
- Occasional swelling.
- Nocturnal symptoms—waking from sleep with numb hands—characteristic of carpal tunnel syndrome.
- Headaches, blurred vision, tearing, tired or burning eyes, difficulty seeing, especially small print.

Figure 2.2 CTD Symptoms

Many of the above symptoms are subjective—only known to the person experiencing the symptoms. Several can be objectively measured. For example, numbness and certain elements of weakness can be determined by special tests known as nerve conduction and electromyography studies. Even pain, perhaps the most subjective of symptoms, can be measured indirectly through various pain measurement instruments.[3]

Symptoms and Conditions

The symptoms and conditions associated with common cumulative trauma disorders are described in this chapter. Additional information on diagnosis and treatment is offered in Chapter 7.

Carpal Tunnel Syndrome

Carpal tunnel syndrome is characterized by numbness and paresthesias in the median nerve distribution of the hand, which is the palm side of the thumb, index and long fingers, and radial side of the ring finger. Nocturnal symptoms are frequent, and individuals may shake their wrists to relieve symptoms—the flick test. More severe carpal tunnel syndrome leads to atrophy of the thenar eminence at the base of the thumb and decreased fine motor movements with loss of thumb abduction and opposition strength. Phalen's maneuver can be used by holding the affected hand with the wrist fully flexed for 60 seconds. Tinel's sign can be elicited by tapping over the carpal tunnel. Both tests are diagnostic if pain or paresthesias occur.

Carpal tunnel syndrome in the workplace has been attributed to compression of the median nerve at the wrist.[4-10] Recent studies suggest that edema, caused by pressure or vibration, of the distal small nerves and glabrous skin receptors is contributory,[11-13] which helps explain negative electrical studies in some carpal tunnel syndrome cases.[14]

de Quervain's Syndrome

de Quervain's syndrome is a tenosynovitis affecting the short extensor and the long abductor tendons of the thumb, and is common in workers who repetitively deviate their wrist in an ulnar and radial direction, such as carpenters, four-part paper users, and butchers. It is also caused by a congenital anomaly in which a tendon in the wrist, extensor pollicis brevis, has its own compartment. With this syndrome, the first extensor compartment is tender. This is the tendon at the base of the thumb. In the Finkelstein test the examiner grasps the base of the patient's thumb and stretches it in ulnar deviation. The test is positive if pain occurs at the radial styloid and along the first extensor sheath. Treatment includes a splint incorporating the thumb. Corticosteroid injection into the first extensor sheath is often beneficial. Surgery may be indicated.

Ulnar Nerve Entrapment

Ulnar nerve entrapment is a variant of carpal tunnel syndrome, with symptoms in the ulnar aspect of the hand and little finger. It is caused by compression or trauma at the medial epicondyle or Guyon's canal. Tinel's sign is positive at either location.

Wrist Tendinitis

It is important to delineate whether hand and wrist complaints are due to median or ulnar nerve entrapment or to tendinitis. The latter is characterized by synovial thickening and pain on resisted wrist maneuvers. In contrast to carpal tunnel syndrome, nocturnal symptoms are infrequent with tendinitis.

Epicondylitis

Epicondylitis occurs with repetitive torquing of the elbow, as in assembly work or construction.

Lateral Epicondylitis. In lateral epicondylitis, or tennis elbow, if the elbow is placed in extension and wrist extension is resisted, pain occurs at the lateral epicondyle and the extensor wad muscles of the forearm. Treatment includes training workers to perform duties with the elbow at 90 degrees of flexion, using a tennis elbow strap, or injecting corticosteroid locally into the lateral epicondyle. Occasionally, surgery is necessary.

Medial Epicondylitis. In medial epicondylitis, or golfer's elbow, tenderness is present over the medial epicondyle area, sometimes called "the funny bone," and pain is elicited by flexing the elbow in 90 percent flexion and resisting flexion at the wrist. Treatment includes localized corticosteroid injections.

Rotator Cuff Tendinitis

Rotator cuff tendinitis, or supraspinatus tendinitis, occurs with work that involves abducting the shoulder with the elbow extended, as with welders and painters for example. The supraspinatus tendon pushes against the acromion in a painful arc or impingement at 70 to 100 degrees of abduction. Treatment includes nonsteroidal anti-inflammatory drugs, corticosteroid injections and physical therapy. Early initiation of range of motion and pendulum exercises can prevent this condition from becoming a frozen shoulder.

Bicipital Tendinitis

Bicipital tendinitis often occurs with supraspinatus tendinitis and is common in workers who reach overhead, such as window washers, construction workers, and stockers. Pain is noted at the anterior shoulder and with movement of the glenohumeral joint. The bicipital groove is usually tender. Treatment is similar to that of supraspinatus tendinitis.

Acromioclavicular Syndrome

Acromioclavicular syndrome occurs in workers who place stress on the acromioclavicular joint while working at waist level, such as construction workers, grinders, and assembly workers. Pain is elicited when the worker pushes downward and during percussion of the clavicle.[15] Treatment includes nonsteroidal anti-inflammatory drugs, corticosteroid injections, and in severe cases, resection of the distal clavicle.

Vibration Syndromes

The use of vibrating tools can lead to an array of symptoms in the upper extremities, including pain, numbness, and dysfunction. Vibration-induced white finger (VWF) is also known as Raynaud's phenomenon of occupational origin. It consists of numbness, loss of muscle control, and reduction of sensitivity to pain and temperature. Hand-arm vibration syndrome (HAV) may mimic cervical or peripheral nerve entrapment of the ulnar or median nerves.[16,17] Treatment involves removal from vibration, the use of splints, nonsteroidal anti-inflammatory drugs, calcium-channel blockers (as with Raynaud's phenomenon), and surgical release of entrapment in unresponsive cases.

Conditions associated with CTDs are summarized in Table 2.1.

Table 2.1 Common Conditions and Features of CTD

Condition	Feature
Neck and Shoulder Girdle	
Cervical syndrome	Pain with movement of neck. Radiates down arm
Tension neck syndrome	Neck pain
Thoracic outlet obstruction syndrome	Variation in pulse with chin forward and hyperextension at shoulder (Adson's test)
Back	
Low back strain	Pain with movement of back (radiating pain down leg if herniated disc)
Upper Extremity (Shoulder)	
Acromioclavicular syndrome	Pain over the acromioclavicular joint when clavicle is tapped while pushing downward against resistance.
Bicipital tendinitis	Pain over bicipital tendon and resisted bicep maneuvers
Frozen shoulder	Decreased range of motion after an injury
Supraspinatus tendinitis (rotator cuff tendinitis)	Pain on abducting the shoulder beyond 70 degrees
Tendinitis, tenosynovitis, bursitis	Local pain, swelling, triggering, catching, crepitus
Upper Extremity (Hands and wrists)	
Carpal tunnel syndrome	Nerve conduction studies: Tinel's sign at carpal tunnel, Phalen's maneuver, the flick test
de Quervain's syndrome	Tenderness of first extensor compartment. Positive Finkelstein test
Upper Extremity (Elbow)	
Lateral or medial epicondylitis	Local pain and pain with resisted hand motion
Ulnar nerve entrapment	Nerve conduction studies, Tinel's sign at ulnar or Guyon's canal

Putting It Together

- The causes of cumulative trauma disorders are many. Use a checklist as a starting point to uncover the causes of CTDs.
- Risk factors for cumulative trauma disorders are often multiple. For example, constrained and repetitive movements at work may occur in combination with a small wrist.
- Look for the obvious. If a wrist bar on a keyboard eliminates hand numbness a cumulative trauma disorder has likely occurred. If this remedy does not eliminate numbness, obtain professional help.

Endnotes

1. Bernard, B., Sauter, S., Fine, M., Petersen, M., and T. Hales. Job task and psychosocial risk factors for work-related musculoskeletal disorders among newspaper employees. *Scandinavian Journal of Work and Environmental Health* 20:417-426, 1994.
2. Salvendy, G., and M.J. Smith. *Machine Pacing and Occupational Stress*. New York: Taylor and Francis, 1981.
3. Sandmark, H., and R. Nissel. Measurement of pain among electricians with neck dysfunction. *Scandinavian Journal Rehabilitation Medicine* 26:203-209, 1994.
4. Armstrong, T.J., and D.B. Chaffin. Carpal tunnel syndrome and selected personal attributes. *Journal of Occupational Medicine* 21:481-486, 1979.
5. Cannon, L.J., Bernacki, E.J., and S.D. Walker. Personal and occupational factors associated with the carpal tunnel syndrome. *Journal of Occupational Medicine* 23:255-258, 1981.
6. Feldman, R.G., Goldman, R., and W.M. Keyserling. Peripheral nerve entrapment syndromes and ergonomic factors. *American Journal of Industrial Medicine* 4:661-681, 1983.
7. Kaplan, P.E. Carpal tunnel syndrome in typists. *Journal of American Medical Association* 250:821-20, 1983.
8. Katz, J.N., and M.H. Liang. Carpal tunnel syndrome and the workplace: Epidemiologic and management issues. *Internal Medicine* 9:66-73, 1988.

9. Masear, V.R., Hayes, J.M., and A.G. Hyde. An industrial cause of carpal tunnel syndrome. *Journal of Hand Surgery*, American Volume. Volume 11, 1986.
10. Silverstein, B.A., Fine, L.J., and T.J. Armstrong. Occupational factors and carpal tunnel syndrome. *American Journal of Industrial Medicine* 11:343-358, 1987.
11. Gelberman, R.H., Szabo, R.M., Williamson, R.V., and M.P. Dinick. Sensibility testing in peripheral nerve compression syndromes: An experimental study in humans. *Journal of Bone Joint Surgery*, American Volume, 65:632-638, 1983.
12. Lundborg, G., Myers, R., and H. Powell. Nerve compression injury and increased endoneurial fluid pressure: A "miniature compartment syndrome." *Journal of Neurology, Neurosurgery and Psychiatry* 46:1119-1124, 1983.
13. Szabo, R.M., Gelberman, R.H., Williamson, R.V., Dollon, A.l., Yarn, N.C., and M.P. Dimick. Vibratory sensory testing in acute peripheral nerve compression. *Journal of Hand Surgery*, American Volume, 9A:104-109, 1984.
14. Grund, A.B. Carpal tunnel decompression in spite of normal electromyography. *Journal of Hand Surgery*, American Volume, 8:348-349, 1983.
15. Guidotti, T. Occupational repetitive strain. *American Journal of Family Practice* 45:585-592, 1992.
16. Olson, N. Diagnostic tests in Raynaud's phenomena in workers exposed to vibration: A comparative study. *British Journal of Industrial Medicine* 45:426-430, 1988.
17. Pyykko, I. Clinical aspects of the hand-arm vibration syndrome: A review. *Scandinavian Journal of Work and Environmental Health* 12:439-437, 1986.

3

What Are the Risk Factors for CTDs?

"It's all in a day's work, the huntsman said as the lion ate him."
—*C. Priestley*

Chapter Objectives

After completing this chapter, the reader will be able to:

- Understand the biomechanical basis for CTDs.
- Recognize risk factors for cumulative trauma disorders, or understand how place, posture, position, and personality affect CTDs.
- Describe how to assess workers' risk.
- Identify signal risks as outlined in the proposed OSHA ergonomics standard.

Case Study

Break Up the Day

Ms. T is a 47-year-old administrative director for an accounting firm. Periodically her boss, who she calls "the procrastinator," assigns her a two-day project with only six hours to do the work. Recently, while pounding a stapler for yet another last minute project due the next morning, she developed deep pain in her right wrist. After multiple unsuccessful treatments and nine months off work, she is learning to cope with chronic wrist pain.

Biomechanics

The basic principles of biomechanics provide an understanding of how the internal stresses placed on the various body parts lead to cumulative trauma disorders. Often there is a direct relationship between work factors, body part maneuvers, and the development of certain cumulative trauma disorders. For example, painters often develop shoulder and rotator cuff problems because of repetitious work above shoulder height. Clerical staff may develop wrist pain due to pressing a pen against four-part or carbonless paper.

These factors of biomechanics are critical to understanding the origin of cumulative trauma disorders in the workplace. Biomechanics includes the following:

- Biomechanical basis for cumulative trauma disorders.
- Biomechanics models and practical applications.
- Body structures and loading related to biomechanics.
- Guidelines for understanding stresses placed on various body parts and avoiding cumulative trauma disorders.

Consider clerical work. The days are gone when office work was considered to be easy. Scientific studies have determined risks of CTDs in the modern workforce in checkout clerks, computer operators, and sewing machine operators. Other occupations have not been adequately evaluated. Clerical and other workers may stress a joint 4,000 to 5,000 times a day (such as a checkout clerk at a grocery store who drags items across a scanner). Piece work and other production rewards also contribute to repetitive joint movement without rest. Work sites can be analyzed for such repetition in other duties and for poorly designed work stations. See Figure 5.3 for a detailed checklist.

Work can be adjusted to the worker's range of motion. Wrist bars for keyboards can be provided to allow wrists to remain in the most favorable neutral position, and lifts and foot rests can be offered. A computer display terminal can be placed so that it is level or slightly below eye level to avoid cervical or neck strain, and the chair can be adjusted so the hips are at 90 degrees of flexion. Bifocal wearers will need to place the screen slightly lower to avoid over-extending the neck. The work site should also be analyzed for humidity, vibration, inadequate lighting, excess heat, and task alteration frequency.

There is nothing new about ergonomic interventions. Saloon keepers knew that patrons would stand longer at the bar if a foot rest, like the brass rail, was placed along the floor. Comfortable drinkers drank more liquor.

Figure 3.1 lists exercises to prevent CTDs from occurring.

> In the case of an average reader, free of a neck or upper extremity condition, it is time to take a reading break and practice a few CTD prevention exercises. The following should be done one at a time.
>
> - Stand up, lean at the waist to each side, straighten then twist, straighten then bend backward.
> - Roll the shoulders.
> - Move the neck up and down.
> - Then move the chin to each shoulder.
> - Move the left ear to the left shoulder and then the right ear to the right shoulder.
> - Dangle arms at sides and gently shake the hands. Then rotate arms at the elbow. Then make a fist. Then relax.
> - Repeat each movement three times. Repeat at regular intervals or purchase software with instructions for the same exercises at regular intervals.

Figure 3.1 CTD Prevention Exercises

Signal Risks

In its draft ergonomics protection standard, OSHA offers a number of terms that, in spite of criticism of the proposed standard, help identify the causes of CTDs.[1]

First, OSHA determined that there are two conditions that lead to further action to prevent or control cumulative trauma disorders: "An employee has daily exposure to one or more of the signal risk factors when performing work tasks; or a work related musculoskeletal disorder has been recorded since the effective date of the standard."

Second, OSHA described "signal risks" as critical to addressing CTDs. What are signal risks? Every job has some risk factor exposures. However, not every workplace presents the same risks for cumulative trauma disorders or musculoskeletal disorders. OSHA indicated that five workplace factors were associated with an increased risk for cumulative trauma disorders or what the agency calls musculoskeletal disorders. OSHA suggests that various musculoskeletal disorders occur to a greater degree when there is more exposure to risk factors during work.

Also, the greater the time of exposure, the more likely such adverse conditions will present themselves. OSHA calls these workplace conditions *signal risk factors* since their occurrence is a "signal of a greater possibility of cumulative trauma disorders or musculoskeletal disorder. In general, risks increase with greater exposure, particularly to a combination of risk factors.[2] Figure 3.2 lists OSHA signal risks.

The signal risk factors include:

- "Performance of the same motion or motion pattern every few seconds for more than two hours at a time." An example is small parts assembly workers manipulating objects in close quarters.

- "A fixed or awkward work posture, for example overhead work, twisted or bent back, bent wrist or kneeling, stooping or squatting for more than a total of two hours." An example is tile layers who kneel most of the day.

- "Use of vibrating or impact tools or equipment for more than a total of two hours." An example is a jack hammer or power sander operator.

- "Forceful exertions for more than a total of two hours," as with a pipe fitter's duties.

- "Unassisted frequent or forceful manual lifting" An example is a shipping department loader.

- "Lifting, lowering, carrying, handling or pushing/pulling animals, people, heavy objects, equipment or tools without assistance from mechanical devices." A stocker is an example.

Figure 3.2 OSHA Signal Risks

OSHA divides manual work into two categories: frequent manual handling and forceful manual lifting. *Frequent manual handling* is defined as: "The number of lifts and the duration of time during which the lifts are performed—for example, manually handling objects more than 25 times in two hours would be frequent manual handling, however, the objects must require ten pounds or more of force. If less than ten pounds of force is needed, this activity would not be considered a signal risk factor."

Forceful manual lifting is: "The amount of force required to handle the item. Manual handling which requires 35 pounds or more of force to perform qualifies as a signal risk factor exposure. Manual handling that requires less than ten pounds of force does not qualify as a signal risk factor exposure. To determine whether handling which requires between 10 and 35 pounds of force qualifies as a signal risk factor exposure includes consideration of both the weight handled and the distance the weight being handled is from the lower back."

The signal risk factors are tools to identify workplaces at higher risk of cumulative trauma disorders. When do signal risk factors not apply? A worker who works in awkward positions may be exposed to a signal risk factor. However, if the same worker had a work station with adequate ergonomic support and could move around or change position on a regular basis then a signal risk factor exposure would be less likely.

How Is Worker Risk Assessed?

Focus on Tasks

Divide work into what tasks are done and how many times a day those tasks are performed. Use the Americans with Disabilities

Act (ADA) division of tasks into *essential* functions, those that are important and done frequently, and *marginal* functions, those that are less important and done less often. For example, a clerk may do word processing most of the day (an essential function) and only rarely answer the phone while a co-worker is at lunch or on a break (a marginal function). See discussion of the Americans with Disabilities Act in Chapter 4 for further information.

Review Workers' Practices

Do workers alternate shifts, share each other's tools, have access to lifting teams, moonlight, or use computers at home, or work in positions where the hands are held in awkward positions? Do they pound keys while others gently tap?

Inquire about Work Tools

Ask workers what they have done to make their job more comfortable? One data entry worker who used her dominant right hand for number entry found that her right wrist pain and little finger numbness improved after she put a pad of Post-it™ Notes under her wrist—an inexpensive, effective alternative to a more costly wrist support or bar.

Ask about Shifts

Do workers use each other's tools or chairs? For example, an emergency room clerk of average stature on the second shift would have to adjust the work station daily after the first shift worker (a large individual) is done and has left for the day.

Focus on Past History

Has the worker had prior CTDs and treatment? An employer can not ask these questions before a job offer because it violates the ADA, as discussed in Chapter 4. Have symptoms improved during breaks on weekends and vacations? Does vibration or cold aggravate symptoms?[3] Ask about prior training for cumulative trauma disorders prevention. Do not ask until the offer of employment has been made (see Chapter 4).

Conduct a Work Survey

- *Computers:* Availability and acceptability of a mouse (some allow the user to alternate fingers), tracking point, touch point, glide point, touch pad, and voice dictation systems.
- *Ergonomic interventions:* Availability of wrist bars, adjustable devices, chairs and table, holders for documents and foot supports.
- *Assembly line:* Loading and packaging tasks, availability of platforms to make work positioning tasks easier.
- *Tools:* Use of off-set handles and pliers that allow the worker to avoid putting the wrist into awkward ulnar deviation and thermoplastic or vibration-dampening handles. Proper size grip or handles. One size does not fit all.
- *Lifting aids:* Scissor lifts with turntables or conveyor tops to make loading and unloading easier. Portable or stationary tilters to put work within better reach; pallet servers (lifts and moves pallets); skid lifters; mechanical lift tables; foot pumps or mechanical scissor combinations (scissor lifts and tilts). Inquire about retrofit of existing work surfaces with ergonomically correct work tops.

Position of Body

- *Neck:* Awkward positions and repetitive bending, including frequent looking up at shelves, bending the neck to the side to hold a telephone receiver, and staring down at a computer display terminal that is too low.
- *Back:* Repetitive bending, twisting, leaning side to side, awkward positions, stooping over at work stations, such as nurses moving a patient in a hospital bed.
- *Lower extremities:* Repetitive bending of the ankle, squatting, working near floor level (such as a carpet layer's duties), and repetitive use of foot pedals.
- *Upper extremities:* Arms and shoulders in repetitive and awkward positions, such as a grinder or painter.
- *Wrists:* Word processors. Check for awkward position of wrists.

Forceful Exertion

Work that requires forceful exertion puts more load on the joints, muscles and tendons. As force increases, more fatigue occurs in the muscles. The force requirement increases with the weight and distribution of the load moved. Check for speed of movement, posture of the worker, degree of friction (as with grinders) and vibration (as with jack hammers).

Time

Repetition decreases the time available for the muscle-tendon unit to rest. Time pressure, increased workload and greater hours of computer use are related to the occurrence of work-related musculoskeletal disorders, according to new research. A study assessed the association of upper extremity musculoskeletal disorders (UEMSD) and work-related factors among employees using video display terminals (VDT) at a large metropolitan newspaper. Forty-one percent of the 973 workers responding to the survey reported some type of UEMSD. Neck symptoms were the most frequently reported (26 percent), followed by hand or wrist (22 percent), shoulder (17 percent), and elbow (10 percent) symptoms. Greater time worked at the VDT was associated with increased hand or wrist symptoms. In addition, increased work-load demands, such as increased time working under a deadline and increased job pressure, were associated with increased neck, shoulder, and hand or wrist disorders.[4]

Vibration

Exposure to vibration occurs in many work situations with the use of power tools and jack hammers, and in certain assembly line work. Hand-wrist vibration syndrome, hypothenar hammer syndrome, vibration-induced white finger, and carpal tunnel syndrome have been associated with localized vibration.[5]

Vibrating tools such as drillers, planers, pneumatic tools, grinders, jack hammers, chain saws, weed cutters, power drills, driver, and other power tools are often implicated. The American Conference of Governmental Industrial Hygienists (ACGIH) has established guidelines indicating that exposure to vibration greater than 12 meters per second squared ($12m/s^2$) should be less than one hour.[6]

Conditions aggravated or caused by vibration include vibration-induced white finger, Raynaud's phenomenon, and hand-arm vibration syndrome (HAVS). Symptoms include numbness and tingling, which may progress to fingers turning pale, most often when it is cold. Unlike carpal tunnel syndrome, individuals with HAVS do not experience nocturnal pain. Antivibration tools are available, although they are not always effective. Simply wrapping a tool with padding is not usually adequate, since workers will squeeze harder to maintain their grip. Treatment includes avoidance, acquiring antivibration tools or, as a last resort, medication. Calcium channel blockers can be used to open up small blood vessels in the hands.

Vibrating controls are also a source of hazard. In contrast to localized vibration present in tool use, whole body vibration occurs in heavy machinery work, and may have some association with back and neck disorders.

Handedness

Left-handed workers may have to adapt to tools, handles, and grips designed for the right dominant majority.

Temperature

Cold temperatures can reduce grip, dexterity and sensitivity in the hands, causing workers to grip tools with greater force.[7] Cold is often associated with wet conditions that compel workers to grasp tools to an even greater degree to maintain an adequate grip, leading to CTDs.

Gender

In a recent study in Ontario, 55 percent of lost time claims for CTDs were among women, who comprised 50 percent of the workers in that province. Why? After work, women go home to activities that may also be repetitive, including ironing, laundry, and folding clothes, which may lead to further cumulative trauma. There are also anatomical differences, such as a smaller wrist diameter, that may contribute to CTDs.[8,9] Other contributing factors may be pregnancy, menopause, and use of oral contraceptives.[10] Work stations and tools have also traditionally been designed for males.

Tilted or Slippery Work Surfaces

In the theater a tilted or angled stage, sometimes called a raked stage, may contribute to cumulative trauma disorders and musculoskeletal disorders. This can be a problem for actors or other stage workers, especially since a steep angle is considered desirable for certain stage productions. One casino boasts it has the world's steepest. Standing on a hard and uneven surface can contribute to back and lower extremity symptoms. Slippery surfaces can lead to twisting of the back and ankles.

Visual Demand

Poor light, bad displays, and glare can contribute to information errors, eye fatigue, and awkward positions. One prevention technique is to turn the visual display terminal off, look for glare from overhead lights, then adjust the screen to reduce that glare.

Organization of Work

Excessive work duration, lack of adequate recovery times, excessive hours, and overtime can contribute to fatigue, reduce the time available for recovery from injury, and may lead to CTDs.

New Work

New workers may not be conditioned to the work demands of an unfamiliar setting and may need to work longer and harder to maintain the same production or work goals of the more experienced worker. A change in the work process for existing workers can contribute to soreness in the muscles, tendons or joints, as can a return from a vacation.

Variability of Tasks

Lack of control or inadequate control over the nature and pace of the work has been shown to lead to physical and other complaints associated with CTDs, and machine pacing has been associated with a higher incidence of musculoskeletal illnesses and absenteeism. Reward systems, such as piece or rate work, may also encourage workers to defer needed rest breaks. Prolonged work leads to a longer recovery period.

Type of Grip

Pinch grips, cased by pressing two fingers together, cause 3 to 4 times more force on tendons then power grips, which are created by the whole hand. Tasks that are repeated every few seconds lead to muscle fatigue and strain. The body can recover if given sufficient time. However, if tendon insertions are stretched without sufficient rest time, in combination with forceful exertions and awkward positions, than cumulative trauma disorders are more likely to occur. "Repetitive" is defined as a cycle time of less than 30 seconds or when one fundamental cycle makes up more than 50 percent of the total cycle. Procedures for screening repetitive work are available.[11]

Contact Stress

A desk edge or other sharp cornered objects can cause increased pressure on a small area of the wrist or other body part and lead to CTD symptoms.

Improve Posture

Standard forms for assessing posture while sitting and frequency of changes in posture can be helpful. Figure 3.3 is a suggested form for assessing posture.

	FORCE	CONTINUOUS EFFORT	EFFORTS/ MINUTE
Neck	_____	_____	_____
Shoulders	_____	_____	_____
Back	_____	_____	_____
Elbow	_____	_____	_____
Wrist/hand	_____	_____	_____
Fingers/legs	_____	_____	_____
Knees	_____	_____	_____
Ankles/feet	_____	_____	_____

FORCE: 1 = Light;   2 = Moderate;   3 = Heavy

CONTINUOUS
  EFFORT: 1 = 6 seconds;   2 = 6-20 seconds;   3 = >20 seconds

EFFORTS/
  MINUTE: 1 = < 1/minute   2 = 1 - 5 /minute   3 = > 5/minute

Figure 3.3 Standard Form for Assessing Posture

Hobbies and Pastimes

Ask about hobbies that might contribute to CTDs, such as gardening, sewing, crocheting, needlepoint, jogging, and participation in sports, such as tennis, racquetball, and handball. Inquire about video game or computer use after work hours. A worker may have an ergonomically correct station at work, but a haphazard setup in a home office. Record what type of bed and couch is used at home. Does the worker slouch in an overstuffed couch without adequate lumbar support for 3 to 4 hours after work?

Control Opportunities

Analyzing workplace conditions requires examining numerous aspects of the job, but a bad situation can be made better if even a few factors are addressed, for example, the proper position of the neck.

- *Personal protective devices* include safety glasses, aprons, and arm and toe guards and should be evaluated. Be cautious about vibration dampening gloves. They can sometimes lead workers to use more force if the gloves do not fit well. Gloves that are too tight can constrict the fingers and inhibit blood flow. Gloves that are too loose can lead to a poor grip and increase the force required.

- *Engineering controls* are defined as measures that change the physical characteristics of the workplace, including equipment, lights, humidity, and temperature. Headsets can reduce neck bending. Wrist rests should have rounded edges and allow the wrist to remain straight. Arm rests can be adjusted or wrapped with foam, and glasses provided with a focal length set for computer screen use.

- *Administrative controls* include procedures that limit exposure to cumulative trauma disorders at work, such as shorter work shifts, rest breaks, and alternating tasks.
- *Substitutions* are product changes and alterations in materials, weight, and dimensions that can reduce CTDs, such as switching to smaller boxes for stockers.

Putting It Together

- Although risk factors for the development of cumulative trauma disorders are many and seemingly confusing, work stations can be systematically assessed by using a checklist with relevant categories of risk.
- Try asking workers "If you could change one thing about your work routine, what would that be?" You will be surprised at the answers and the increased openness. You might also request staff to think through a possible solution in advance if they offer a problem.
- The proposed OSHA ergonomics standard, though not in effect, offers an action plan based on the above risks and their severity or signal risks that can guide cumulative trauma disorder prevention and control efforts.

Endnotes

1. CFR 1910. OSHA Proposed Ergonomics Standard, 1995.
2. See note 1.
3. Kaji, H., Honma, H., Vsui, M., Yasuno, Y., and K. Saito. Hypothenar hammer syndrome in workers occupationally exposed to vibrating tools. *Journal of Hand Surgery*, British Journal, 18:761-766, December 1993.
4. Bernard, B., Sauter, S., Fine, L., Petersen, M., and T. Hales. Job task and psychosocial risk factors for work-related musculoskeletal disorders among newspaper employees. *Scandinavian Journal of Work and Environmental Health* 20:417-426, 1994.
5. See note 3.
6. 1993 *Threshold Limit Values for Chemical Substances and Physical Agents and Biological Exposure Indices*. Cincinnati, Ohio: American Conference of Governmental Industrial Hygienists, 1993.
7. See note 6.
8. Ashbury,. F. Physical work load and pregnancy outcome. *Journal of Occupational Environmental Medicine* 37:941-944, 1995.
9. Armstrong, T.J. Carpal tunnel syndrome and the female worker. *In:* Transactions of the forty-third annual meeting of the American Conference of Governmental Industrial Hygienists. Portland, Oregon, 1981.
10. Cannon, L.J., Bernacki, E.J., and S.D. Walter. Personal and occupational factors associated with carpal tunnel syndrome. *Journal of Occupational Medicine* 23:255-258, 1981.
11. Gilad, I. A methodology for functional analysis in repetitive work. *International Journal of Industrial Ergonomics* 15:91-101,1995.

Part II
CTD Prevention and Control

"But of good leaders, who talk little, when their work is finished, their aim fulfilled, the others will say 'We did it ourselves'."
—Lao Tzu

Part II encompasses the five P's: purpose, plan, process, partners, and persistence.

Chapter 4:	Fitting the Worker to the Work	. 59
Chapter 5:	Evaluating the Demands of the Job	79
Chapter 6:	How to Fit Work to the Worker	117
Chapter 7:	Diagnosis, Prevention, and Treatment of Common CTD's	. 137
Chapter 8:	Symptom Prevention	175

4

Fitting the Worker to the Work

"The significant problems we face cannot be solved at the same level we were at when we created them."

—*Albert Einstein*

Chapter Objectives

After completing this chapter, the reader will understand:

- The risks, liabilities, and current issues regarding:
  - Employee selection programs, such as preemployment, preplacement, and postoffer examinations, as a means of decreasing CTDs.
  - The Americans with Disabilities Act: what are reasonable accommodations for workers with CTDs, and the essential and marginal functions of a job.
- How to intervene appropriately in work station tasks and practices to reduce cumulative trauma disorders.

- How the self-responsibility model for managers and employees and the benefits of involving the employee as a partner can reduce cumulative trauma disorders.

Case Study

The Right Tools for the Right Job

Mr. J. is a 34-year-old plumber who, while turning a large three-inch rusted pipe, realized a larger tool would be more effective. He decided to continue without it and yanked harder and dislocated his right shoulder. He has been plagued by chronic pain since.

How does one fit the worker to the work, in contrast to the reverse, which is reviewed in Chapter 6? Approaches to prevent cumulative trauma disorders that involve fitting the worker to their work include:

- Worker factors.
- Worker selection programs, such as strength testing and postoffer examinations.
- Worker training.

Worker Selection Programs

Worker selection is often defined as a means of fitting the worker to the work—or, as the Marines suggest, finding a "few good men." Understanding the demands of the workplace is critical. Do not just settle for job titles. For example, employees at a glove factory include hookers, who take freshly made gloves off forms with a hook and strippers, who strip gloves off forms. Ask for details to avoid misunderstandings!

Extrinsic versus Intrinsic Factors

Employee selection programs focus on the extrinsic and intrinsic factors associated with cumulative trauma disorders. The extrinsic or the physical demands of a particular job may produce certain injuries. Of particular concern are jobs involving twisting, carrying, pulling, lowering, bending, lifting, torquing, awkward work position, and unexpected movements. Combinations of certain of these activities can be particularly hazardous, including lifting and twisting at the same time.

Understanding the unique characteristics of each job is critical. After the job is understood, different approaches to preventing cumulative trauma disorders, including changes in work training, work practices and techniques, and job design or ergonomics can be implemented. Job redesign has great potential for preventing cumulative trauma disorders. Ways to analyze a job from an ergonomics standpoint and determine the risk of injury are reviewed in Chapter 6.

Worker Training

Knowledge of the extrinsic and intrinsic factors of tasks can assist in developing appropriate worker training. Intrinsic factors include worker strength, fitness, size, and height, as well as any preexisting conditions that might affect cumulative trauma disorders and their occurrence, including diabetes, pregnancy, arthritis, and gout. An example of an extrinsic factor is proper lifting. In the past, workers have been instructed to use their legs to lift while keeping the back straight to avoid bending forward with rotation, which increases lumbar disc pressure 20-fold when carrying 10 kg.[1] Recent research findings advise that we analyze the extrinsic factors of the job. If the load is too large to be brought between the legs, or if it is located horizontally away from the feet, bending at the waist may be preferable to using the legs because intradiscal pressures are the same when the ratio of moment arm to load is constant.[2]

In general, workers should be instructed to hold heavy loads close to the body and to avoid sudden twisting of the waist. They should be reminded that low back pain occurs not only from lifting, but also from push and pull activities and from working on slippery and uneven surfaces.

Workers whose activities are physically challenging should be informed that they are "industrial athletes." They should warm up before work and perform conditioning exercises at least three times a week to prevent injury. Readily available instruction sheets should be offered.

Cautions on Worker Selection Programs

Worker selection programs should be considered with caution, with a clear understanding of their purpose, and awareness of government regulations, such as the Americans with Disabilities Act (ADA). The ADA offers challenges and opportunities for management and human resources professionals. These will be discussed later in this chapter. Although a worker's strength, range of motion, and endurance can be measured, in general such job selection programs are poor predictors of future CTDs.[3-5]

Preplacement or Postoffer Exams

First a word about terminology. "Preemployment" testing was the favored term before the ADA was instated. Under the ADA, when a conditional job offer is made, then a medical examination or other testing occurs, hence the use of the terms "preplacement" or "postoffer."

If an applicant is not selected for work because of a medical condition, a medical examiner must demonstrate that a prospective employee's prior medical condition poses a direct threat to the safety and well-being of others, that no reasonable accommodation can be made, and that significant objective findings with proven predictive value are evident. A recent innovation that may be of help in ADA accommodation issues is the functional capacity evaluation, which analyzes strength and range of motion and matches the physical capabilities of the injured, recovered, or disabled worker with expected duties. See Chapter 5 for further information.

Strength Testing

Job-specific strength testing is a controversial means of identifying workers at higher risk for CTDs, as are preplacement radiographs and inquiries about prior low back pain. These have limited predictive value.[6] In a well-designed study, Snook concluded that worker selection based on strength testing showed no correlation with the risk of subsequent injury.[7]

Preplacement (preemployment or postoffer) strength testing also has poor predictive value. Research at a steel mill revealed no difference in the incidence of back injuries between women workers who met the preemployment lifting criteria and those who did not. In 230 applicants for heavy labor positions, no difference was detected between lifting and trunk strain testing in those who later developed a low back injury and those who did not.[8]

X-Rays

Radiologic studies have no role in preplacement examinations, except when they are justified by a clinical examination or a disease state such as spondylolisthesis.[9]

Electrodiagnostic Studies

What about preplacement electrodiagnostic studies as a method of evaluating the peripheral nervous system and identifying individuals prone to carpal and ulnar tunnel syndrome? One study tested nerve conduction levels at the beginning of employment. Electromyographic (EMG) gaps of the trapezius muscle during work were studied as a risk for CTD. "Contraction" levels were significantly higher in those who later received cumulative trauma disorder treatment than those who did not.[10] Although this study is suggestive, electrodiagnostic studies should never be done in isolation from an adequate history and physical examination. Additional research needs to be conducted before such studies can be relied upon in a preplacement setting. Proceed cautiously. Job applicants have filed lawsuits after they allegedly failed a preplacement nerve conduction evaluation for CTDs.

Figure 4.1 offers several suggestions for what can be done to provide a better fit between workers and their work.

> - Self-responsibility and the worker as a partner.
>
>   Workers support what they are invested in. Offer workers opportunities to join the CTD-prevention team. Their suggestions will be helpful and their support invaluable. It will take more time, but it is time well spent. Work attitudes are also critical in preventing CTDs. One study found that as workers grew older they were less positive about their ability to prevent low back pain.[11] Managers are advised to assist workers in developing a proactive prevention attitude early, especially as the workforce ages.
>
> - Mark packages and containers with their approximate weight or contents so the worker can gauge the strength required to lift that item.
>
> - Consider warm-up exercises before a work shift. One mine safety officer initiated such a program and injuries declined seventy-five percent.[12]
>
> - Strength can be increased. The U. S. military has provided women with strength training that enabled them to meet the physical requirements expected of men, such as carrying a 40-pound pack.

Figure 4.1 Fitting the Worker to the Work

Americans with Disabilities Act

On July 26, 1994, all U.S. employers with 15 or more employees were required to comply with the Acts' Title One on employment, one of the most readable government documents written, but also one of the more controversial. Even with the publication of the Equal Employment Opportunity Commission's final rule and technical assistance booklet, many of the provisions will be settled in the courts. Preplacement, postoffer, and return-to-work examinations and related record keeping are under new restrictions, and health professionals and others are more involved in reasonable job accommodations. For example, an amplified stethoscope can be provided for a hearing-impaired nurse.

The challenge of the ADA lies in practicing in a manner consistent with the requirements, providing qualified disabled workers with appropriate work and ensuring their safety and that of others. Cumulative trauma disorders are generally covered by the Act and employers are increasingly being called to task under the Act, though there are notable exceptions. Recently, a grocery store worker with a CTD was not considered sufficiently disabled under the ADA.[13]

Accommodations

Cost of accommodations under the ADA range from no cost to expensive. The average cost of the accommodations in 1994 were:

- No cost — 18 percent

- $500.00 or less — 50 percent

- $1,000.00 or less — 10 percent. More than $1,000.00 — 22 percent[14]

For further information about job accommodations for CTDs, call the Job Accommodations Network at 304-293-7186.

If an employer fails to make a reasonable accommodation, a worker with a cumulative trauma disorder may file a discrimination case with the Equal Employment Opportunity Commission. The Commission heard 208 cumulative trauma disorders cases in 1994.

A sample post-offer/preplacement medical evaluation form including restrictions and accommodations is presented in Figure 4.2. A sample physical capacities classification form is presented in Figure 4.3.

CONFIDENTIAL
Post-Offer/Preplacement Medical Evaluation

Employee name:_____ Date:_____

Employer:_____

Job title:_____

Based upon my evaluation:

_____ 1. No medical contraindication to performing this job.

_____ 2. No medical contraindication to performing the essential functions of this job, with the following restrictions and/or accommodations:

_____

_____

_____ 3. Based upon probability of substantial harm, this employee could pose a direct threat to self or others. The extent of the threat and accommodations which may decrease the threat:

_____

_____

_____ 4. Medical hold: waiting for additional data. Re-evaluate on ___ /___/___

_____ 5. Further testing is required to fully evaluate ability or risk.

Comments:_____

_____

_____
Healthcare Provider

Figure 4.2 Post-Offer/Preplacement Medical Evaluation

PHYSICAL CAPACITIES CLASSIFICATION

NAME (LAST, FIRST, MIDDLE)	SS NUMBER	DATE

POSITION	WORK LOCATION

TO HUMAN RESOURCES	FROM (HEALTH CARE PROVIDER)

The remaining portion of this form to be completed by the Health Care Provider:

I. QUALIFICATIONS:

☐ NO RESTRICTIONS
☐ RECOMMENDED WITH RESTRICTIONS LISTED BELOW

☐ RECOMMENDED FOR JOB EXAMINED FOR ONLY
☐ OTHER

II. RESTRICTIONS: CHECK AS MANY AS NEEDED TO IDENTIFY WORK RESTRICTIONS.

AMBULATORY
☐ No excess walking or standing
☐ No working on elevators
☐ No excess stooping/crouching or bending
☐ No overhead work
☐ No climbing stairs or ladders
☐ Other _____

IRRITANTS
☐ Avoid irritating oils and coolants
☐ Avoid solvents or greases
☐ Avoid soaps or detergents
☐ Avoid irritating chemicals
☐ Other allergens, sensitizers or irritants specify) _____

LIFTING
☐ No pushing or pulling
☐ Not to lift, push or pull over _____ pounds
☐ No carrying or lifting
☐ Other _____

OTHER
☐ Avoid manual labor
☐ Avoid heavy manual labor

Keyboarding:
☐ 15 ☐ 20 ☐ 30 ☐ 45 minutes
with breaks of:
☐ 5 ☐ 10 ☐ 15 ☐ 30 minutes
Total keyboarding time/day:
☐ 1 ☐ 2 ☐ 4 ☐ 6 ☐ 8 hours
☐ No keyboarding or typing

ENVIRONMENTAL
- ☐ Avoid extreme heat and/or cold
- ☐ Avoid wet or damp areas
- ☐ Avoid dust and fumes
- ☐ Avoid excessive sun exposure
- ☐ Avoid work in unpressurized aircraft in flight or altitudes above 8,500
- ☐ No shift work
- ☐ No overtime work
- ☐ No outdoor work
- ☐ No offshore work

MACHINERY
- ☐ May not operate or work near moving and/or hazardous machinery
- ☐ May not drive crane or motor vehicle
- ☐ May not use vibratory tools

HEARING-EARS
- ☐ Must wear ear protection in identified noisy environment
- ☐ Must wear hearing aid on job
- ☐ Avoid job requiring non-perforated eardrums (toxic gas exposure)
- ☐ Avoid job requiring good hearing
- ☐ Other_____

RESPIRATORY-RESPIRATORS
- ☐ Qualified for use
- ☐ Qualified for emergency use only
- ☐ Not qualified
- ☐ Other_____

VISION
- ☐ Must wear visual correction on job
- ☐ Must work in well lighted area
- ☐ Must wear safety glasses on job
- ☐ Avoid job requiring color sensitivity
- ☐ Avoid job requiring good near vision
- ☐ Avoid job requiring good far vision
- ☐ Avoid job requiring good depth perception
- ☐ Avoid job requiring good lateral vision
- ☐ Avoid job requiring good vision in both eyes
- ☐ Other_____

RESULT TO:_____
DATE:_____
TIME:_____

REPORTED BY:_____

Comments:
_____
_____
_____
_____
_____
_____

RETURN TO DUTY (if applicable) _____
DURATION OF RESTRICTION _____

PROVIDER NAME AND SIGNATURE:

_____
SIGNATURE   (PRINT NAME)   DATE

Figure 4.3 (cont'd) Physical Capabilities Form

Impact of the ADA

Irving R. Tabershaw, M.D., former American College of Occupational and Environmental Medicine (ACOEM), president, in a letter to Les Shaptini, M.D., past president of ACOEM described the impact of the ADA:[15]

"I think this law will have the most serious impact on the practice of occupational medicine since the OSHA law was passed in 1970. We should develop a consensus on how to function professionally and responsibly under the ADA mandate and the regulations promulgated by the EEOC."

Specific concerns which Dr. Tabershaw identified included:

- "The content, scope, tests, procedures, etc. of the preplacement examinations have in the past been arbitrarily adapted by the company or a contract physician without any peer review. Most of the protocols have never been made public and may be open to attack as discriminatory, unfair, biased, incompetent, and not pertinent to the job. If litigation develops, they will be made public.

- Job descriptions, if available at all, are inadequate and, for the most part, unsuitable to determine a valid physical and medical job criteria.

- The examining physician has a "tall" order to advise the employer on the accommodations he would make to overcome the applicant's or worker's handicap. As occupational physicians, we have had no experience with this task. Yet, the employer will have to rely heavily on the physician who performs the medical fitness examinations to advise him properly."

Ideally the ADA should:

- Help integrate the disabled worker with a CTD into the workforce.
- Match worker abilities with the essential functions of a job.
- Guide employers to discuss reasonable job accommodations for workers with CTDs.

Figure 4.4 outlines key compliance considerations of the ADA. The ADA flow chart in Figure 4.5 covers the critical action points in the job application process as they relate to CTDs. Figure 4.6 is a sample job description including essential and marginal functions.

What Can One Ask about a Worker's Disability under the ADA?

At the pre-job offer stage you can ask job applicants and current employees whose health has changed:

- Can you perform the essential job functions?
- Can you demonstrate how you would do the job and/or complete any part of the hiring process (typing test, interview, demonstration of job)?

1. Does a disability exist (either a true disability, a history of disability, or simply perceived as a disability)?

    For example: A worker who cannot write because of a CTD may be disabled, but one who cannot write because of dropping out of school is not.

2. Does the disability substantially limit one or more major life activities?

    For example: A worker sustained a CTD that permanently restricted lifting ability. This is a disability. Another worker, also with a CTD, was able to continue working. This is not a disability.

3. Temporary disabilities do not qualify.

    For example: A carpal tunnel syndrome that heals is not a disability. A carpal tunnel syndrome that leads to median nerve damage and causes permanent numbness in the hand is a disability.

4. Essential Functions: Can the disabled applicant or worker perform essential functions unaided or with reasonable accommodation?

    First, ask whether employees in the position are actually required to perform the function. For example: An office job description says the person files, but the essential functions are to answer the telephone and filing is handled by others. An individual impaired by a CTD who cannot file, but who can answer the telephone should not be considered unqualified.

    Second, would removing that function fundamentally change the job?. For example: A company needs a supervisor to cover when regular supervisors are absent, and this is the only reason this position exists. It is therefore an essential function.

    Third, essential functions should focus on outcomes. For example: A job that requires objects to be moved from one place to another should state this as an essential function. The job description might say "lifts 40-pound boxes four hours/day," but should not say "manually lifts 40 lbs" as an essential function unless manual lifting is the only method possible without causing an undue hardship to the employer. A mechanical lift might be an accommodation.

Figure 4.4 Considerations for Compliance with the Americans with Disabilities Act

5. Does a direct threat to the safety of the employee or others exist based on objective medical evidence that cannot be reduced with a reasonable accommodation? Direct threat is defined as a significant, specific, and current risk.

   For example: A cook who has an active hand infection poses a risk to the health of others. However, with medication and a brief absence from work, the cook would not pose a direct threat. The brief absence is a reasonable accommodation.

   What about a nerve conduction test to find CTD-prone individuals, those in the early stages of carpal tunnel syndrome? First, the test must be predictive. Second, the test must be related to an essential function. Third, the abnormality on such a test must pose a direct threat that cannot be reduced with an accommodation.

6. Preplacement (postoffer or preemployment) physicals: Decisions must be job related. A position may be withheld when no reasonable accommodation is possible.

   For example, if a medical history reveals a worker has suffered serious multiple CTDs doing similar work that have progressively worsened the condition, employing this worker would incur significant risks of reinjury.

7. Accommodations are unreasonable if costly, extensive, substantial, or if they fundamentally alter the nature of the workplace.

   For example, a visually impaired theater worker wanted the lights turned up in a theater with subdued lighting. The employer may not have to comply because doing so would fundamentally change the nature of the establishment.

8. What to do when an employee is having difficulty performing the job? In such cases a medical examination may be necessary to determine if the employee can perform the essential functions with or without an accommodation.

   For example, in the case of an employee who has erratic behavior and recurrent absenteeism and performance problems, a medical examination may be necessary to determine if a medical condition is the cause. If not, the employee does not have a disability and an accommodation is not required. The ADA does not protect a worker with a bad attitude.

Table 4.4 (cont'd) Considerations for Compliance with the Americans with Disabilities Act

AMERICANS WITH DISABILITIES ACT, TITLE I

Effective Date - 07/26/92

JOB APPLICATION PROCESS - ACTION POINTS FOR EMPLOYERS

Job Applicant	Disability	Interview/Ref. Check/Drug Screen	Qualified Conditional Job Offer	Accomodations	Injury/Illness--Workers' Comp. vs Off-Work	Return to Work

- Essential Functions of Job
- Demo. of Functions by Applicant
- Medical Inquiry/Exam
- Permanent Impairment from Injury/Illness

Definition of Disability:

1. Physical/mental impairment which substantially limits 1 or more life activities
2. Record being disabled.
3. Regarded as disabled.

Requirements for Exam:

1. Exam must be for everyone in that position.
2. Exam must focus on essential function.

Permanent Impairment from Injury/Illness

1. Look at essential functions of job.
2. Can employer accommodate additional impairment?

Rules for Accomodations:

1. Pose no direct threat to self or others.
2. Impose no undue hardship upon employer (financial, etc.).
3. Call Job Network for information (1-800-JAN-7234)

Figure 4.5 ADA Flow Chart

Summary: Assists in computer-aided instruction. Operates and manages the site network system.

Essential Functions: Operates networked system, manages users and groups on file server. Provides technical information to assist in planning and managing the delivery of instruction. Assists in the planning of developmentally appropriate lessons to meet instructional objectives. Assists with computer lessons to ensure comprehension of content. Monitors progress in the computer lab. Plans, prepares, and conducts training sessions on use of computers. Researches new software that may enhance computer-assisted learning. Maintains equipment in the computer learning lab. Troubleshoots computer operations problems. Arranges for the repair of computers.

Marginal Functions: Schedules use of lab and prepares materials for lab use. Maintains inventory records of hardware, software, and lab materials. Prepares warehouse requisitions for computer lab supplies. Participates in site committees to plan effective use of instructional technology.

Mental Tasks: Communicating, reading, comprehending. Ability to understand written and oral instructions. Requires observing the behavior of students in the classroom.

Physical Tasks: Requires ability to operate network computer system. Listens, writes. Requires manual dexterity. Requires hand-eye coordination.

Methods, Techniques, Procedures: Ability to instruct in a group setting.

Equipment, Aids, Tools, Materials: Operates computer equipment. Utilizes manuals and forms.

Working Conditions: Indoors. Classroom environment. Contact with children, parents, employees.

Supervision, Control: None

Minimum Qualifications: Verbal and written communication skills. Two (2) years of computer lab experience. Evidence of basic math principles and/or use. Knowledge of computer equipment and its uses. Any equivalent combination of training, education, or experience that meets the minimum qualifications.

Figure 4.6 Sample Job Description (Essential and Marginal Functions)

If the individual asks for a reasonable accommodation, the employer must provide it or ask the applicant to describe how they would perform the job with such an accommodation.

- Can you provide documentation of your disability? This may be asked if the applicant requests a reasonable accommodation.

During the interview or on the job application, do not ask:

- If the applicant has a disability.
- If the applicant needs reasonable accommodations.

At the postjob offer stage:

- The employer can send the individual for a medical evaluation as long as all individuals for that job classification or position are required to do so.[16]
- The medical evaluator may inquire about disability, and may make recommendations about reasonable accommodations, as long as they are limited to the functions of the job.[17]

Putting It Together

- Fitting the worker to the work is fraught with liabilities and risks. Strength and other testing programs have been found to be poorly predictive of future CTDs.
- For those employers with 15 or more employees, adherence to the Americans with Disabilities Act is the law. Specific terms, such as essential functions and marginal functions should be memorized.

- Under ADA restrictions, never ask a job applicant or employee returning from an illness or injury if they have a disability. Employers can ask the employee or the applicant to demonstrate the essential functions of a job. Only after the position has been offered can employers inquire about what reasonable accommodations are necessary to perform the essential functions.

Endnotes

1. Rueler, J.B. Low back pain. *Western Journal of Medicine* 143:259-265, 1985.
2. Lavender, S.A., and G.B.J. Andersson. Ergonomic principles applied to the lumbar spine. *Journal of Disability* 3:1-15, 1993.
3. Himmelstein, J.S., and G.B.J. Andersson. Low Back pain: risk evaluation and prereplacement screening. *Occupational Medicine* 3:255-268, April-June 1988.
4. Snook, S.H., Campanelli, R.A., and J.W. Hart. A study of three preventive approaches to low back injury. *Journal of Occupational Medicine* 20:478-481, 1978.
5. Dueker, J.A., Ritchie, S.M., Knox, T.J., and S.J. Rose. Isokinetic trunk testing and employment. *Journal of Occupational Medicine* Volume 36, 1994.
6. See note 3.
7. See note 4.
8. See note 5.
9. American College of Radiology: Conference on low-back x-ray's in preemployment physical examinations. *In:* Proceedings of meeting, Tucson, Arizona, January 11-14, 1973.
10. Veiersted, K.B., Westgaard, R.H., and P. Anderson. Electromyographic evaluation of muscular work pattern as a protector of trapezius myalgia. *Scandinavian Journal of Work and Environmental Health* 19:284-290, 1993.
11. Hyytiainen, K. Attitudes toward prevention of low back disorders in industry. *Occupational Medicine* 44: May, 1994.
12. P. Maley, ASARCO, Tucson, Arizona, Personal Communication, 1989.

13. Taylor vs. Albertsons, Inc., CA 10, No.95-6112, 1/11/95.
14. Job Accommodation Network, 1994. Telephone (304) 293-7186.
15. Tabershaw, Irving. *The ACOEM ADA Handbook: The Americans with Disabilities Act—Challenge and Opportunity for Occupational Health and Human Resource Professionals.* Arlington Heights, IL: American College of Occupational and Environmental Medicine, 1992.
16. Pruitt, R.H. Preplacement evaluation: Thriving within the ADA guidelines. *AAOHN Journal 43 (3):124-130,1995.*
17. *The Americans with Disabilities Act: Your responsibilities as an employer. EEOC-BK-17 (800-669-3362).*

5

Evaluating the Demands of the Job: Worksite Functional Job Analysis

"What we need is a technology of human behavior."
—B. F. Skinner

Chapter Objectives

At the end of this chapter, the reader will:

- Understand the key elements of a functional job analysis.
- Know how to evaluate the physical demands of a particular job.
- Be able to collect the essential information about the job process and job tasks.
- Be aware of the revised NIOSH lifting guides.
- Understand what an ergonomically correct computer workstation should look like.

- Be able to analyze computer workstations to identify risk factors.
- Be able to modify workstations to improve their ergonomics.
- Understand why some back injuries are being recognized as a cumulative trauma disorder.
- Be aware of what functional capacity evaluations are and their use.

Case Study

Teamwork

An assembly plant is growing rapidly and would like to consider ergonomic principles in their work processes in order to reduce the increasing incidence of CTD. One department where tubing is placed on plastic outlet valves has already had several upper extremity injuries. Twelve percent of the employees in a second department where 25-pound cases are placed on pallets and tipped have a history of low back pain. The plant manager has asked that an evaluation of the jobs in these departments be conducted and has established a team to assist in problem solving and implementing changes.

Worksite Functional Job Analysis

Worksite functional job analysis techniques consist of evaluating the functions of a job and its demands, promoting work team problem solving, collecting information about the functions of the job, both marginal and essential, relating function to the work capacity of the worker, and predicting cumulative trauma disorders in the work environment. These elements are all critical to preventing and controlling cumulative trauma disorders.

A functional analysis to evaluate the demands of the job and the NIOSH revised lifting guides are tools to cost-effective problem solving and can guide problem-solving teams in preventing and controlling cumulative trauma disorders.

An organized problem-solving approach to evaluating the demands of a job and revising them as needed can be conducted by the following steps:

- Select team members to collect information about the job's demands, both essential and marginal. See Chapter 4 for an explanation of the difference.
- Determine who your partners are. Remember the worker has the most intimate knowledge of the job and should be included in the process.
- Diagram the work flow and station.
- Establish what the purpose is; for example, to decrease CTDs.
- Then determine the plan. Review the data and identify the areas of highest incidence of CTDs.
- Work out the process. Prioritize high CTD areas based on the feasibility of making changes.
- After these steps have been established the various job demands can be related to the work of the individual. Cumulative trauma disorders can be predicted for high risk tasks, and problem solving can be facilitated.

Worksite Design and Ergonomic Assessments

Ergonomic assessments often focus on the loads imposed on a worker's joints and posture, but ignore the dynamic motion of these joints, which can lead to CTDs. More modern ergonomics assessments allow a dynamic evaluation, including the effect of motion on joints and the loading of joints. Dynamic ergonomic assessments increase understanding of the workplace and the effects of joint motion on cumulative trauma disorders.

The revised NIOSH lifting guide recognizes the importance of dynamic assessment and can help determine safe weight limits for manual lifting. This determination can be made manually or by use of computer software. The revised guide and a list of software vendors are listed in Figure 5.1.

The Revised NIOSH Lifting Guide

The revised NIOSH lifting guide offers a method of evaluating manual lowering and lifting work and analyzing the relative risk of low back injuries and overexertion.[1,2] There are two key components to the guide—the lifting index and the recommended weight limit—which can help determine if a job is safe.

Lifting Index (LI)

The lifting index (LI) is the relative risk of back injury and overexertion for a given lifting task. LI equals load weight/recommended weight limit. The lower the LI, the safer the job. In general, an LI less than or equal to 1 (one) is acceptable.

Recommended Weight Limit (RWL)

Under a given set of conditions, the load that 75 percent of female and 99 percent of male workers can lift safely is the RWL. The equation
$$RWL = LC \times HM \times VM \times DM \times AM \times CM \times FM$$
is used to evaluate the recommended weight limit in a given task.

Load Constant (LC)

If the object to be lifted is greater than 51 pounds, the task may not be safe for certain workers.

Horizontal Multiplier (HM)

The HM is the horizontal distance from the ankle to middle finger knuckle—avoid lifts when distance is greater than 25 inches.

Vertical Multiplier (VM)

The VM is the distance of hands above the ground/floor.

Distance Multiplier (DM)

The DM is the travel distance between the beginning and end of the lift using the hands as the measuring point.

Asymmetric Multiplier (AM)

The asymmetry line is the line joining the midpoint of the hands projected to the floor and the line between the ankles. The asymmetric angle is the angle the asymmetric line makes from the mid sagittal plane of the body—the line drawn down the middle of the body's vertical axis.

Coupling Multiplier (CM)

CM is a measure of the ease of handling parts and objects.

Frequency Multiplier (FM)

Frequency is the number of lifts per minute.

Strain Index

The strain index is a means of measuring the exertion of the distal upper extremity (wrist, hands and fingers). A score greater than five is considered unsafe, over seven is hazardous.[3]

Revised Guide
Revised NIOSH Guide Program for Manual Lifting, A. Garg Industrial & Manufacturing Engineering University of Wisconsin-Milwaukee P.O. Box 784 Milwaukee, WI 53201 414-229-6240
Software Vendors
Three Dimensional Static Strength Prediction Program University of Michigan Software Program Intellectual Properties Office The University of Michigan Ann Arbor, MI 48109 313-764-8202

Figure 5.1 Revised Guide and Software Vendors

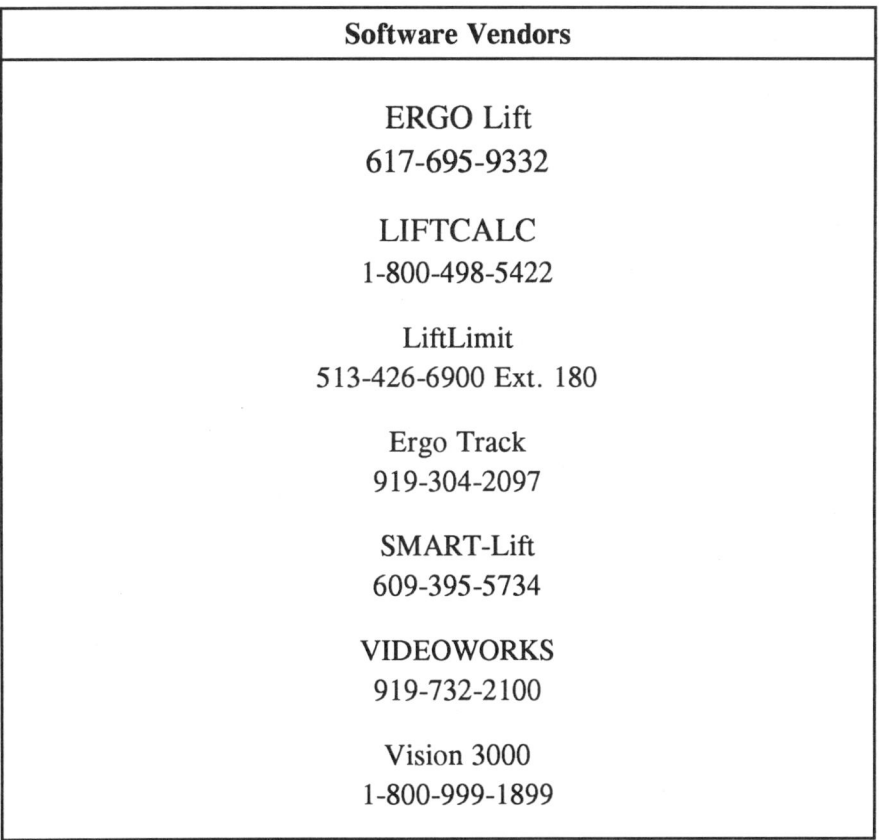

Figure 5.1 (cont'd) Revised Guide and Software Vendors

In order to understand when a station poses a high risk for CTDs, the principles of low risk or proper workstation arrangement should be understood. Computer workstations are one of the common environments conducive to CTDs. Because workers have different physical characteristics, computer work areas need to be adapted to the individual. Figure 5.2 illustrates an ergonomically correct computer workstation. Be aware that even in the ergonomically perfect workstation, workers wiggle and move around at regular intervals, and so they should, since the human body was designed to move, not sit still for eight hours a day.

Figure 5.2. The feet are well supported. Hips and knees are bent to approximately 90 degrees. The back is well supported. The wrists are supported and in a neutral position, and the neck is in a neutral positon.

The following should be adhered to:

- Feet should be flat on the floor with equal weight on each foot.

- Hips and knees should be bent to 90 degrees.

- The back should exhibit its three postural curves and be well supported, as in Figure 5.2.

- The wrists, and, if possible, the forearms and elbows, should be well supported.

- The wrists should be in a neutral position, not bent up, down or sideways.

- The neck should be in a neutral posture, not bent up or down or sideways.

These are general guidelines. When dealing with CTDs there is not one solution for every situation. Each setting needs to be evaluated separately to accurately identify the risk factors. When this has occurred, recommendations can be made to try to decrease the risk factors. If these can be decreased sufficiently, symptoms will diminish, be relieved, or never occur at all.

Figure 5.3 offers a checklist to aid in analyzing computer workstations. Figure 5.4 illustrates points covered by the checklist.

Department/Employee: _____

Analyst: _____ Date: _____

Ergonomic Keyboard Assessment

☐ Use of a foot rest ☐ Use of a wrist rest

☐ Wrists in neutral position, ☐ Neutral head position
elbows at 90° flexion or >

Posture/Support/Practices Criteria	Acceptable	Action
1. Monitor Height		
Just below eye level	(Y) (N)	_____
Lower for bifocal workers	(Y) (N)	_____
2. Monitor Distance		
18 - 21	(Y) (N)	_____
Glare	(Y) (N)	_____
3. Sitting Posture		
Straight	(Y) (N)	_____
Supported	(Y) (N)	_____
Angle trunk & thigh 90°	(Y) (N)	_____
4. Wrist Position		
Straight	(Y) (N)	_____
Supported	(Y) (N)	_____
Are forearms parallel to floor?	(Y) (N)	_____

Figure 5.3 Checklist for Computer/Clerical Workstation

5.	Feet			
	Supported	(Y)	(N)	_____
	Equal weight bearing	(Y)	(N)	_____
6.	Legs			
	Angle of thigh & calf greater than 60°	(Y)	(N)	_____
7.	Are documents at eye level	(Y)	(N)	_____
8.	Mouse accessible	(Y)	(N)	_____
9.	Breaks			
	Worker takes occasional, small breaks when working at the computer for long periods	(Y)	(N)	_____
	Stretching exercises at the work station	(Y)	(N)	_____
10.	Adjustable chair & keyboard	(Y)	(N)	_____
11.	Often-used items easy-to-reach	(Y)	(N)	_____
12.	Awkward posture, excessive repetition, contact with sharp edges, or excessive force	(Y)	(N)	_____
13.	Task Rotation	(Y)	(N)	_____
14.	Personal Protective Devices	(Y)	(N)	_____
	Splints	(Y)	(N)	_____
	Wrist rests	(Y)	(N)	_____
	Back supports	(Y)	(N)	_____

Figure 5.3 (cont'd) Checklist for Computer/Clerical Workstation

90 / CUMULATIVE TRAUMA DISORDERS

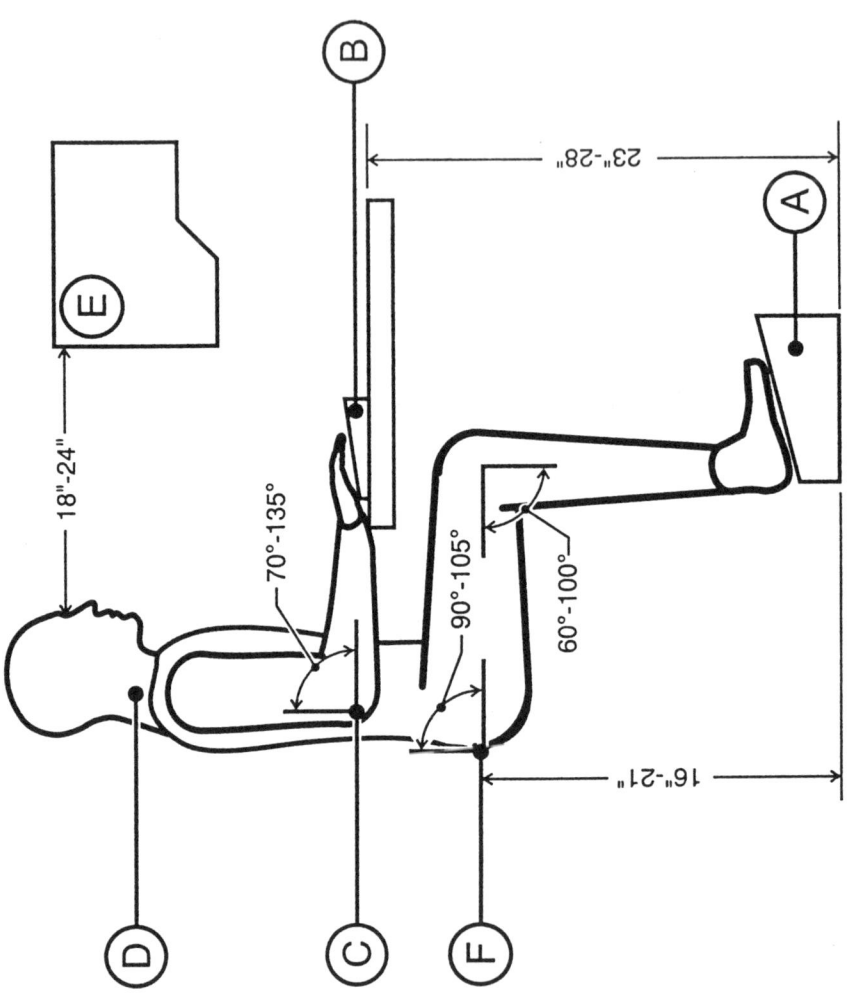

Figure 5.4. Diagram of ideal computer workstation.

Problem Solving

Figure 5.5 shows an incorrect computer workstation. The worker's wrists are bent down (flexed). The feet are off the floor, and the back is poorly supported.

Foot Supports

Many of us sit at our desks without having adequate foot support (Figure 5.6). Ideally, the feet should be flat on the floor or on a foot support with equal weight on each foot (Figure 5.7a and 5.7b). Common problems in the workplace, include:

- When short workers adjust their chairs so their wrists and neck are in a good posture with good foot support, their feet may not touch the floor. A foot rest can alleviate this situation.

- Resting the feet on the base of the chair or putting one leg up on the chair and sitting on it are two habits of workers that may contribute to poor posture and increase the risk of developing a CTD.

Many risk factors are the result of worker behaviors and are the activities most difficult to change.

Figure 5.5. The primary problem here is that the work surface is too low. Because of this, the chair also needs to be lowered so the worker can fit under the work surface. As a result, the knees are higher than the hips and hit the work surface; the feet are poorly supported and not flat; the sitting posture is poor and the wrists are too extended. In addition, there is no wrist or upper extremity support. Finally, the monitor is too high and too far away, causing neck extension.

Figure 5.6. A "worker behavior." The employee is choosing to sit in this manner, which does not provide good foot support.

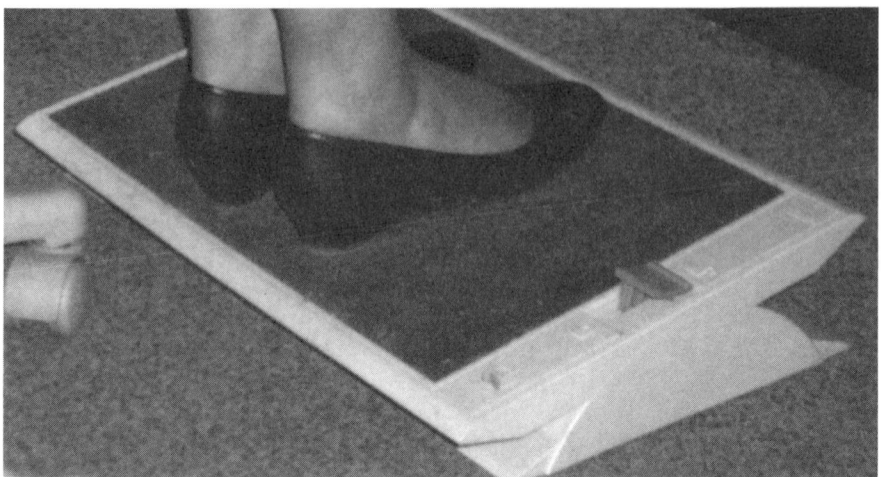

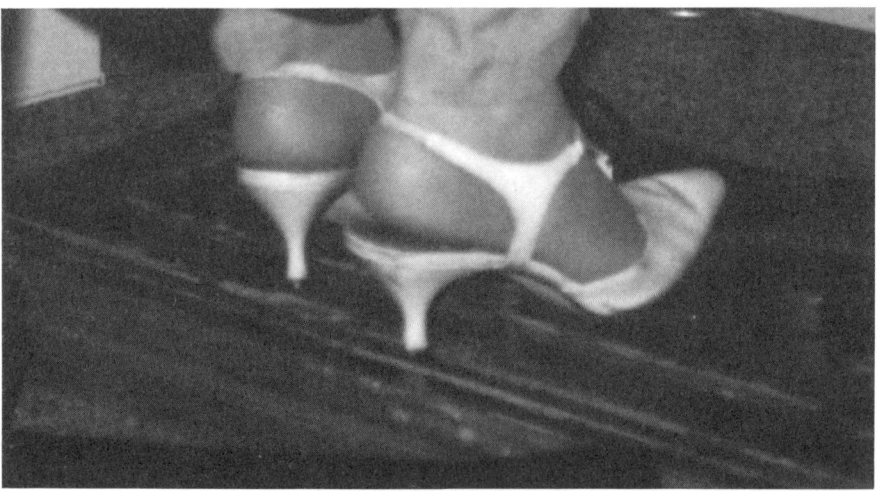

Figure 5.7a and 5.7b. Proper foot support. Figure 5.7a is a $45 foot rest. Figure 5.7b was brought from home and is made out of a few pieces of wood.

Hips and Knees

While sitting, hips and knees should be at approximately 90 degrees, which provides good support of the back. If employees are tall or wear high-heeled shoes, this recommendation may be difficult to follow if their knees hit the edge of the desk or a drawer. Those with certain types of back problems may prefer to be in a position where their knee is below the hip rather than horizontal at 90 degrees, a position that increases the angle between the thigh and the trunk and may relieve stress, pressure, or pain in the back.

Back Support

The back should be well supported with either a well fitting chair, cushion, or both. Avoid sitting in a slouched posture (Figure 5.8). Sitting up straight in a more neutral posture puts less stress on the back (Figure 5.9).

Wrists and Arms

The wrists should be supported on either the work surface, a well-fitting wrist rest or both (Figure 5.10). The more the upper extremity is supported, the better. Well-fitting arm rests on a chair can provide additional support for the forearms. The wrists should be kept in a neutral position both in the up and down plane and in the side to side plane (Figure 5.11 and Figure 5.12).

96 / CUMULATIVE TRAUMA DISORDERS

Figure 5.8. A slouched posture. There is one big C curve instead of the three natural curves in the back and neck.

Figure 5.9. Good back support along with good foot support and monitor height. The left wrist is well supported, but is in slight extension rather than a neutral position.

98 / CUMULATIVE TRAUMA DISORDERS

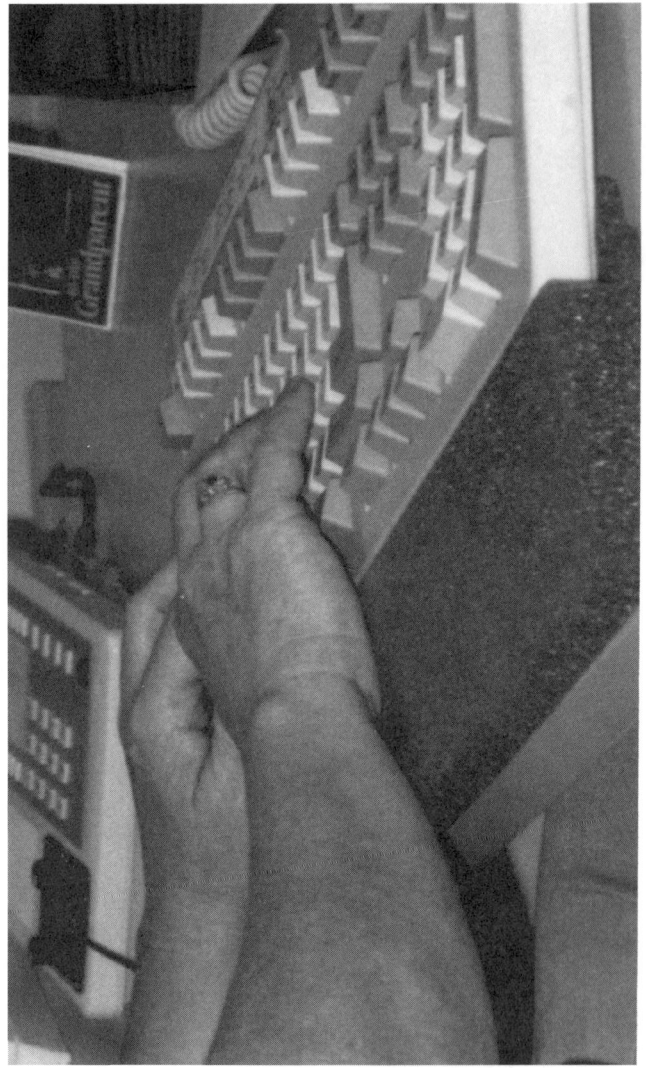

Figure 5.10. Good support and positioning of the wrists. The supports on the back of the keyboard were collapsed (folded in) to create a flatter angle of the keyboard, which avoids putting the wrists in an extended position.

EVALUATING THE DEMANDS OF THE JOB: WORKSITE FUNCTIONAL JOB ANALYSIS / 99

Figure 5.11. The left and right wrist are extended and turned inward. Both of these positions should be avoided.

Figure 5.12. For most people, the top of the screen should be at or slightly below eye level to avoid constantly looking downward. However, if you wear bifocals like this employee does, the screen should be lowered in order to provide for the right focus. If the screen was at eye level, she would have to tip her head back in order to see the screen through the bottom portion of her lenses.

Figure 5.13. Often, if the screen is too far away, the worker will lean forward to get closer. This can affect back and neck posture and support.

102 / Cumulative Trauma Disorders

Figure 5.14. A monitor that is too close can affect posture and support as well as vision.

Head, Neck, and Shoulders

Stress in the neck muscles can occur if the computer monitor is too low and the worker must constantly look down at the screen. In general, the screen should be at or slightly below eye level. If bifocals are worn, the screen needs to be lowered (Figure 5.12). Workers wearing bifocals look through the bottom portion of the lenses. If the screen is at or slightly below eye level, their necks must be extended to see the screen through the lower part of their lenses. Just as repeatedly looking down at the screen can cause stress on the neck, so can repeatedly looking up.

The screen should be 18 to 24 inches away from the eyes. Frequently, it is either too far away (Figure 5.13) or too close (Figure 5.14). Glare on the monitor screen can be another problem. This may be caused by overhead lighting, natural lighting, and the location of the screen. An anti-glare screen can often cut down on this problem. Changing the location of the monitor in relation to artifical or natural lighting coming through a window can also help. Turn off the monitor, observe where glare strikes the screen, and then either adjust the monitor location, decrease overhead lights or use window shades to reduce the glare or add an anti-glare monitor filter. Add a task light to improve illumination of documents.

Completing the CTD Puzzle

Figure 5.4 shows an easy-to-follow diagram that can be used to start analyzing a work area. When modifying a work area, try to get all of the pieces to fit like a puzzle. As you use this tool to try to decrease CTD risk factors, you will find some of the ideas will work better than others.

As one risk factor is decreased or eliminated, another risk factor may be made worse or a new risk factor may be created. Assessing and modifying a work area often involves trial and error. When a chair is raised to obtain better alignment of the wrists, the monitor height is altered. When the chair is raised to the correct height for the monitor, the worker's feet may no longer touch the floor.

Keep in mind that the principles presented here are basic. Many professionals specialize in performing this kind of work. They can be called upon when a situation is beyond your knowledge or comfort level. These professionals may include, but are not limited to, ergonomists, engineers, industrial hygienists, physical therapists, occupational therapists, physicians, and nurses. Ergonomics is a specialty, therefore not everyone in these disciplines has the necessary knowledge and experience to address problems associated with CTDs. Inquire about the professional's background and always ask for references.

Assessing the Work

To accurately analyze a worker's work site, inquire about what it is that they do. Ask the worker to make a sketch or complete a list. By learning what the worker's tasks are, it is easier to determine what is physically necessary to carry out the job tasks. This information is then assessed for risk factors, such as high repetition rate, long duration, and heavy force.

Lifting is one of the important factors to consider when looking at what a worker does. What is being lifted? How much does it weigh? How often is it lifted or tilted? Where is it lifted to and from? Is it a one-person lift or are others required? Is an assistive device, such as a forklift, available? These questions should also be asked for carrying, pushing, or pulling. The type of flooring should be considered. Is it uneven, steep, or slippery? Is anti-fatigue matting (cushioning) available? A risk factor that can easily be decreased is faulty or poorly maintained carts or other equipment that is pushed or pulled.

The various positions of the workers and their duration need to be considered. These positions include sitting, standing, walking, climbing, bending, twisting, kneeling, and reaching.

Assessing the Worker

A critical element in assessing work is to observe the worker performing job tasks, and then assessing or reassessing the risk factors for CTDs. Remember, a worker's habits can play a large role in the development of CTDs. Does the worker have the strength, range of motion, and coordination to do the job? Is movement done in a smooth flowing manner? Do movement patterns or techniques used to perform the work put the worker at risk for a CTD? One worker developed arm pain from continually reaching above shoulder height for a reference book. When the book was moved to a lower shelf the pain resolved. Look at posture. Does the worker choose poor posture, or does the nature of the work or the work site cause poor posture? Is the worker bending when squatting is the preferred posture? Is the worker choosing to twist or rotate instead of moving the whole body—a maneuver that can contribute to a back injury. Instead of "working smart," is the worker making the job harder physically by assuming a position with a mechanical disadvantage?

Assessing the Worksite

After analyzing the work and the worker, the worksite should be assessed. This may include evaluating:

- amount of space

- layout of space
- lighting (is glare a factor?)
- organization of the desk (how far does the worker have to reach for tools?)
- work surface (is it slippery or uneven?)
- storage of material and supplies (are they too high or low or inaccessible?)
- vibration characteristics, temperatures, and humidity
- availability of assistive devices (lifts, headsets, automatic staplers)
- presence of protective devices (finger guards, etc.)
- sharing of tools and equipment, including chairs, with co-workers

Space

Space is usually expensive and companies often offer the minimum amount possible per worker—a risk factor that cannot be controlled. Conversely, where there is a little extra space, workers tend to spread out to mark and protect their territory. This presents another risk. Items placed farther away cause workers to reach a greater distance more frequently—a factor that could lead to CTDs.

Layout

The work space layout should be set up like a kitchen, laundry area, or workshop at home. Items more frequently used are usually placed between knee and shoulder height. Less frequently used items are most often on the top and bottom shelves because they are harder to get to and require greater effort and strength. The work station should follow a similar pattern. A problem-solving technique is to condense an individual's work area so that the frequently used items are within arm's length. The layout of the work space should also take into consideration the amount of twisting, turning, or rotating done by the worker. Something as seemingly innocuous as the placement of a telephone can add a tremendous amount of turning to daily work. It may not seem important, but observe how many times a receptionist reaches for the phone to either answer it or to make a call in just one hour? Then multiply that frequency by eight hours in a day and forty hours in a week and the repetitions are considerable. If the telephone is placed in a location that requires both reaching and turning, the worker combines two different risk factors, often occurring at a high frequency.

Chairs

Ergonomically designed chairs may include enhanced lumbar support, seat depth and back height adjustment, arm and elbow rests, arm width adjustments, and a floating seat to keep the user's feet on the floor. Additional scientific evidence to support the benefits of such equipment as a means for decreasing the incidence of CTDs is needed, but their use may decrease certain CTD risk factors, increase worker satisfaction, and improve productivity.[4]

Keyboards

Revised keyboards are an attempt to reduce fatigue and limit excessive wrist and finger movements—both considered causes of CTDs. Some examples of revised keyboards are:

- Split keyboard: Especially effective for individuals with large hands or those with wrists bent outward, such as an ulnar drift in rheumatoid arthritis.

- Keybowl: Two domes are moved laterally on an alphanumeric input device that limits flexion and extension of the wrist and finger movement.

- Revised layout: Changing the key layout can decrease finger movement. The most popular keyboard or QWERTY was actually developed to slow down manual typists who typed so fast that the older manual typewriters became stuck. The Dvorak and the As In Red Hot are attempts to place the most commonly used keys on the home row (see Table 5.1).

Table 5.1 Keyboard Layout

Name	Home Row
QWERTY	asdfghjkl
Dvorak	aoeuidhtns
As In Red Hot	asdeihotnr

Lighting

The lighting in a work area may create an environment that is too bright or too dim. Either situation can lead to symptoms such as headaches, blurred vision and difficulty seeing. The lighting may affect workers' posture because of placing themselves in a different or awkward position in order to see better. Glare particularly affects computer users. Changing the location of the computer monitor, if that is possible, or using an antiglare screen can help.

Organization

The organization of the space, supplies, and the storage of material can be addressed in a fashion similar to the layout of the work space. The key is to organize efficiently. Put items together that are used together or that tend to be replenished at similar times. One worker with wrist pain caused by pulling five separate sheets from five separate folders to create a new patient file found that if she used the copier's automatic collator to group the sheets, she had less discomfort.

Nonphysical Issues

Other issues that may play a role in the development of CTDs include morale, conflict with supervisors, and layoffs. Often a worker is unhappy with some aspect of life, whether it be work related or not. A work-related injury may force them to be off work or to be placed in another department doing modified duty that might cause a CTD. One worker with a low back injury was placed in a sitting job without being able to change position—a position that aggravated his back pain.

These nonphysical issues need to be identified and addressed, or at least acknowledged, otherwise improvement is less likely. Explain to the worker that if a car is taken to a mechanic and the mechanic cannot figure out what is wrong, the problem is unlikely to be fixed. If a patient goes to a doctor and says one thing is making him sick when it is really something else, what is prescribed to correct the problem probably will not work. For example, a marital problem might not be solvable, but once it is recognized the right resources can be offered, such as counseling or an employee assistance program.

Low Cost or No Cost Modifications

In an era of corporate cut backs, downsizing and reengineering, many organizations do not have money budgeted for modifications of the workplace. On the other hand, if a modification or accommodation will prevent a CTD and allow the worker to continue working in a productive manner, providing the modification may provide a cost benefit.

The first approach to consider in solving a CTD problem should be a low cost or no cost one. For example, instead of purchasing a wrist rest, a rolled up towel, foam, sponges, a rubber pad, a Post-it™ Notes pad or the cardboard spool from a paper towel roll may suffice.

Instead of buying a foot rest, a telephone book can be recycled to accomplish the same result. It can be covered with duct tape to increase durability. Table 5.2 lists ergonomic interventions and low cost alternatives.

Table 5.2 Low Cost Approaches to Ergonomic Intervention

Ergonomic Intervention	Low Cost Alternative
Lumbar pillow	Towel rolled up and taped
Wrist rest	Towel Paper towel roll Post-it™ Notes Foam or rubber pad
Foot rest	Telephone book Wood blocks Masonry blocks
Pencil or pen grip	Tape

The Cumulative Nature of CTDs and Back Injuries

A typical history: Eighty to 90 percent of adults will suffer from back pain.[5] For most the pain will resolve in a few days without the need for medical treatment. The next time the pain occurs, it may last longer and require more rest, heat, ice, or even medication to bring about a recovery. More time may elapse, and once again pain recurs. This time it may last longer than before and a visit to a doctor is required. Remedies might be a prescription for physical therapy, medication, or a back brace. In time recovery occurs, only to have the pain return in a few months. The interval between episodes of pain may shorten and the duration of symptoms may lengthen—a typical history of a cumulative trauma disorder, whether it be in the wrist or in the back.

It is important to understand the history of CTDs and back pain. Workers have presented to health care providers stating: "All I did was bend over to pick up a magazine off a coffee table and my back went out." But it was not that one-time event of lifting an object that weighed a few pounds that caused the back injuries. In reality it was the hundreds of times that they have bent over incorrectly in the past. It is difficult to understand that these cumulative back injuries can occur in the absence of pain until an episode such as the one described above occurs and becomes "the straw that broke the camel's back."

Prevention

Explain to workers that the spine is composed of vertebrae stacked one on top of the other like building blocks. Between each pair of vertebrae is a disc. The disc has two parts: an inner part and an outer part. Describe the inner part as being similar to the jelly inside of a doughnut and the outer part, the annulus fibrosis, as a roll of toilet paper. When a person bends forward at the waist and the back becomes rounded, pressure is exerted on the disc. Recall what happens to the jelly inside the doughnut when one bites into it. By compressing the front of the doughnut, pressure is put on the jelly. If this is done enough times or with enough force, the jelly will push out of the back of the doughnut. The result is a protruded (bulging) or extruded (herniated) disc.

Every time a person bends over, the jelly part of the disc is pushed out, putting pressure on the inside ring of that roll of toilet paper. After enough repetitions or with enough force, the jelly will break through that first ring. If the bending continues, other rings may be broken through as well. If the jelly gets through most of those rings, the disc itself can change shape and start bulging. If the process continues, a piece of the disc ruptures and a piece may break loose. This is called a herniated, uptured, or extruded disc. The nerves that enervate various parts of the body exit the spinal column on the sides of the disc. If the bulge is large enough it may compress the nerve, causing a "pinched nerve." This can be painful and sometimes requires surgery to trim or remove part of the disc.

Caution the worker not to bend over to write a quick note at a desk or bend over to pick up a pen. The correct procedures are to sit down to write or to get down on one knee to pick up a pen from the floor. This will prevent the pain, suffering, and other adverse effects that accompany a back injury. Although it might sound ridiculous, it is well worth adopting this method.

Twisting is another motion that is hard on the discs. Twisting creates a shearing type of force that puts stress and pressure on the disc. By moving the feet with the trunk of the body, twisting at the waist is avoided. Table 5.3 reviews other CTD prevention techniques.

Table 5.3 Work Techniques to Prevent CTDs

Action or Risk Factor	Recommended Technique
Mopping—often done by twisting at the waist.	Use the legs to move side to side holding the mop in front.
Vacuuming—as above or by bending at the waist.	Use the legs to move backward and forward holding vacuum cleaner and arm steady.
Working under vehicle hood—often done by bending over at the waist without support of the back (same for brushing teeth).	Use one arm as a prop—holding on to the vehicle (or sink) to support the back.
Reaching overhead.	Use a step stool.
Heavy lifting.	Break up the task by alternating with lighter tasks.

Functional Capacity Evaluations

What are functional capacity evaluations and how can they prevent and control CTDs? A Functional Capacity Evaluation (FCE) is an assessment of what a worker can and cannot do from a functional and physical perspective. It defines what someone is physically capable of performing. If a worker is not able to perform an activity, the FCE can often provide the reason. This is known as the limiting factor.[6]

An FCE can include testing any physical aspect of work. Typical items assessed include lifting, carrying, pushing, pulling, sitting, standing, walking, reaching, hopping, sliding, fingering, manipulating, handling, bending, twisting, crouching, kneeling, and squatting. Depending on the type of work, other items such as climbing and crawling can also be added. The FCE may require one or two or more days. The longer the assessment, the more accurate the information. A second day of testing also allows the evaluator an opportunity to assess how the work was tolerated after the first day of the testing.

Points to Consider

- *Caution versus overdoing the evaluation.* The former may not provide enough information. The latter, if excessive, may harm the worker being evaluated.
- *Consistency.* It is important to evaluate consistency of effort, the lack of which may indicate poor compliance with the testing.
- *Flexibility.* Each FCE should be individualized. Every worker is different, as are their symptoms.
- *Prevention.* An FCE may also assist in preventing recurrence of injuries if the person has surgery and returns to the same work setting. The symptoms may return unless a revision has been made. Also an FCE can assist in rehabilitation so that a weak body part is not subject to the same work station without return of muscle tone and mobility.

Putting It Together

- Evaluating the demands of the job using a functional job analysis offers a tool to cost-effective CTD problem solving. Other tools include the revised NIOSH lifting guide, ergonomically correct workstations, and understanding the demands of the job, all of which can guide problem-solving teams in preventing and controlling CTDs.

guide problem-solving teams in preventing and controlling CTDs.

- Back injuries can usually be considered another form of cumulative trauma. An exception is back injury from a motor vehicle accident or a fall. These CTDs are equally amenable to the prevention strategies for other CTDs.

Endnotes

1. Waters, T.R., Putz-Anderson, V., and A. Garg. *Applications Manual for the Revised NIOSH Lifting Equation*. DHHS (NIOSH) Publication No. 94-110. Cincinnati, OH: National Technical Information Service (NTIS) PB94-176930, 1994.
2. Waters, T.R., Putz-Anderson, V., Garg, A., and L.J. Fine. Revised NIOSH equation for the design and evaluation of manual lifting tasks. *Ergonomics* 36(7):749-776,1993.
3. Moore, J.S., and A. Garg. A comparison of different approaches for ergonomic job evaluation for predicting risk of upper extremities disorders. *In:* Proceedings of 12th Triennial Congress of the International Ergonomics Association. Toronto, 1994.
4. *CTD News*: Dec 10, 1994.
5. Peate, W.F. Occupational musculoskeletal disorders. *In:* Campos-Outcalt D, ed. *Occupational Medicine: Primary Care*. Philadelphia: WB Saunders, 1994.
6. Rodgers, S.H. A functional job analysis technique. *Occupational Medicine: State of the Art Reviews* 7(4):674-711, 1992.

6

How to Fit Work to the Worker

"The best way to predict the future is to create it."
—*Peter Drucker*

Chapter Objectives

After completing this chapter, the reader will:

- Know the basics of strategies to control CTDs for the upper and lower extremities and in manual handling and lifting.

- Know how improving ergonomics can decrease the occurrence of CTDs.

- Understand the key elements of an ergonomically correct work station.

- Recognize that ergonomics is a young science, as well as an art, and that knowledge is constantly being added or revised.

- Be aware of new recommendations regarding visual display terminals and other computer components.

Case Study

On the Home Front

A new employee complains of numb hands, which wake her from sleep and affect her crochet work. She thinks her keyboard at work is the cause and wants the new "Fashion and Function" workstation that she saw advertised at the mall. What should you do?

Work Site Design and Ergonomics

What is ergonomics? Ergonomics is designing the job or fitting the job to the worker. Some workers' jobs involve activities and motions that their bodies cannot sustain. This leads to pain, injury, illness, and cumulative trauma disorders. Ergonomics is more than muscles and skeletal conditions. Ergonomics attempts to find a better match between a worker's physical capabilities and limitations and the work activities and conditions of the job through improved operation and design of equipment, controls, work stations, and tools. The aim is to help workers work at their best.

Walk the floor. Ergonomics works best when it involves a multidisciplinary team. Healthcare providers and safety professionals should play a key role. Rather than treat five individuals from the same department with similar cumulative trauma disorders, physicians and others are advised to visit the workplace to investigate the cause of cumulative trauma disorders and prevent it in others. These walk-throughs are one form of ergonomic assessment.

There are numerous methods of enhancing ergonomic assessments and measurements. Some are:

- *Biomechanics.* Biomechanics is a method of measuring the various external stressors involved in lifting and repetitive work. Such measurements are becoming increasingly quantitative and sophisticated, including attaching sensors to the worker's body to record the amount of repetition, number of posture changes, and duration of forceful exertion.

- *Production factors.* Production area factors involved in CTDs include job rotation, length of shift, and overtime. Heavier work may require short shifts with more frequent breaks or periods of lighter work. The schedules of recovery and work depend on the demands of the job. Overtime often contributes to excessive fatigue and may need to be limited.

- *Job rotation.* Allow workers to rotate through different job activities in order to avoid fatigue in a single body part.

- *Job content.* Enlarge the job by scheduling a variety of different tasks for each worker, a strategy which allows workers to use different muscles rather than one muscle repetitively without rest. A good example is the production of the IBM ThinkPad ™ notebook computer. The computers are produced in Scotland where one worker combines all component parts for a single machine, rather than having a group of workers on an assembly line each handle a single part.

- *Organization.* High demands and little control can lead to physical discomfort. Individuals often at risk are midlevel managers who must implement the demands of top management, but who may have little input into decision making.

- *Gender.* Many work surfaces were designed for men, not women, based on Air Force data. Tables may be too high, chair pads too long, overhead reach distance too far, and handles and grips too thick for women to use comfortably.[1] Factors such as pregnancy contribute to increased physical workload. Prolonged walking and standing may also have an adverse effect. Work guidelines for pregnancy should be reviewed with the individual's health care provider.

- *Worker variety.* Workers are varied. They do not fit every premade workstation. One size workstation does not fit every worker. Workers vary in their arm and leg length, hand size, body width, vision, height, and range of motion. Smaller employees may be unable to reach the work. Larger employees may find there is not enough room to reach into openings. A shift worker may share a work space with another different size co-worker. Consider the following:
    - Design for variety. When possible, purchase or create adjustable work surfaces to allow the smallest to the largest worker to do their job effectively. An example of an adjustable work surface is a computer terminal that swivels in height and distance from the worker.
    - Consider the various extremes in size. Design the work surface height so it can be moved from the 95 percentile height male to the 5 percentile height female. Allow for adequate clearance for body parts so proper posture and reach is allowed. Make adjustments for the extra height and width of shoes, hard hats, and steel-toed shoes as needed.
- *Start with dimensions,* such as height and distance to the computer display terminal. Then focus on actions, such as twisting and reaching. Figure 6.1 provides a form to use to record dimensions of a work space.

More than one strategy can be employed at a time. Table 6.1 outlines such categories and control strategies.

Dimensions	Height/Length/Size	Action
Fit or clearance		
Reach to a shelf or bin		
Height (distance to work level):		
	Length	**Action**
To computer terminal		
Seat length		
To floor from seat		
Angles	**Degree**	**Action**
Neck to trunk		
Trunk to hip		
Elbows to body		
Wrists to forearms		
Diameters	**Diameter**	**Action**
Grips		
Handles		

Figure 6.1 Dimensions and Angles of Work Surfaces Form

Table 6.1 Control Strategies

Categories	Strategies
Extremities:	
Upper	Reduce repetition
	Alter force
	Improve posture
	Decrease contact stress
Lower	Alternate standing
	Avoid squatting/bending/crouching/ stooping
	Revise sitting
Hand tools/gloves	Use proper gloves
	Improve grip design
Vibration:	
Whole body	Decrease vibration transmission
Localized	Balance tools
Work organization	Reduce bending/twisting/pushing pulling/carrying
Manual handling/lifting	Revise reaching
Lifting	Reduce weight

Prevention Strategies for Upper Extremities

The following list reviews CTD prevention strategies for the upper extremity.[2-6]

Reduce Repetition

- Work enlargement: increase the number of different tasks performed by each worker, or rotate employees between jobs that require different muscle groups.

- Provide mechanical assists. (lifts, turntables, etc.)
- Use multifunction tools that allow the worker to alternate body parts used.
- Change the product or the process. One surgical glove manufacturer altered the process of pulling gloves off forms so that workers could avoid flexing their wrists.
- Allow time to rest or stretch. Have new workers start at 50 percent of production goals the first two weeks and allow adaptation.

Alter Force Required

- Change the size or shape of objects that are held in the hands so that they are more manageable.
- Increase the friction of the object in the hand through the selection and texture of handle and glove materials, and by keeping handles free of moisture, grease, and oil. Reduce the weight of hand-held objects by picking up fewer objects at a time, by lifting with both hands or by means of a mechanical assist.
- Grasp objects with a power grip using the whole hand, rather than with a pinch grip using two fingers.
- Grasp objects at their center of gravity so that their weight does not twist them out of the hand.
- Shift the center of gravity by reducing, shifting, or adding weight to one side of the object.
- Balance tools so they are less stressful on one hand.
- Use air shutoff or external torque bars to control tool torque.

- Use mechanical assists for turning, holding, and lifting work.
- When possible slide parts, rather than lift. Use roller or power conveyors and chutes or turntables.
- Use handles that are long enough to be gripped to avoid pinching.
- Replace or service dull and worn tools, casters, rollers, and conveyors. Post a maintenance schedule and adhere to it.
- Avoid gloves that are too bulky or tight. Have different sizes of gloves available, especially if a splint or bandage must be worn underneath.
- Cover only those parts of the hand that must be covered. Use safety tape for fingers or use fingerless gloves, use palm pads to protect palms and leave the fingers free.

Improve Posture

- Design tasks so the work can be performed with the elbows close to the side of the body, so that the forearms and wrists are not excessively rotated, deviated, flexed, or fully extended.
- Allow for frequent change of position.
- Control posture through the location, shape, size, and orientation of the work.

Decrease Contact Stress

- Increase the size and length of handles.
- Wrap handles with tape or thermoplastic materials.
- Use malleable or compliant materials.

- Cover arm rests with foam.
- Eliminate or pad sharp edges that come into contact with body parts to avoid nerve compression. A keyboard wrist bar with rounded edges is an example.

Prevention Strategies for Lower Extremities

What are CTD control strategies for the lower extremity?[7]

Alternate Options for Standing

- Standing stationary for long periods can lead to discomfort in the legs and lower back. Jobs should be designed to allow workers alternate options for sitting and standing.
- A sit/stand chair allows workers with high work surfaces to lean against a firmer surface and to stand with greater ease (These workers often sit on stools with poor back support.)
- A cushioned surface to stand on and foot rests.
- Well-cushioned shoes for workers who stand continuously.
- Avoid using foot pedals when standing. This maneuver leads to awkward hip and back movements.

Avoid Kneeling/Squatting/Crouching/Stooping

- Provide a cushioned surface, such as knee pads or padding on the floor.
- Alternate tasks that require kneeling with tasks that do not.

- Avoid repetitive squatting, kneeling, crouching, and stooping. If possible, limit these to short durations.

- Avoid awkward positions when foot pedal controls are used. When the foot remains in contact with the pedal or requires frequent repetitions, use a pedal that is level with the floor when activated or use a seated work station. Pedals should be three inches or longer.

Revise Sitting

- Alternate sitting with other positions. Take stretch and stand breaks. Use a sit/stand chair at high work surfaces.

- Allow for adequate lumbar and foot support. Use a lumbar pillow to maintain the normal "S" shape of the back.

Evaluation of Hand Tools and Gloves

Power Grip Design

- Avoid tools that cause contact stress by excessively pressing into soft tissues of the hand, causing pressure on skin, tendons, bones, nerves, and blood vessels.

- Tools should not dig into the palms. Use handles that extend along the whole hand.

- Avoid sharp edges that press into the fingers or palm, as with thin clipper handles.

- Avoid handles with finger grooves that add extra pressure on the hands. The groves do not fit every hand size.

- Use handles that allow the entire hand and fingers to remain on the grip.
- Obtain handles with finger stops at the base. These allow for better tool control and also decrease the amount of force needed.
- Use thermoplastic, friction tape, or other material to cover handles.
- Avoid bending the wrists when gripping tools. Keep the wrists straight. If possible, bend the tool, not the wrist.

Improve Force Characteristics

- The weight of a tool and its operation influence the force required to grip the tool. The heavier the tool the more grip force and muscle fatigue occur.
- Limit torque. Use mechanical aids such as clutch-type, shutoff, or hydraulic pulse tools, or those mounted on articulating arms.
- Balance the tool's weight about the grip axis or use tools that weigh under 5 pounds (2.3 kg) if the tool's center of gravity is far from the employee's wrist.
- Use lighter tools for fine precision work, such as in dentistry and electronic assembly work, to reduce muscle fatigue.
- Use side or auxiliary handles to balance tools.
- Avoid tasks that raise, extend, or flex the elbow repetitively.
- Suspend heavy tools with a balancer, arm cords, or hoses.
- Use tools that either hand can hold.

Use Gloves

- Gloves protect workers from friction, trauma, chemicals, cold, heat, and abrasives.
- Choose gloves carefully for proper fit and wearability.
- Avoid gloves that are too stiff or too loose.
- Use a glove with a wrist closure or drawstring if debris can fall into the glove, as in a ditch diggers' work.
- Gloves should be snug but not tight to avoid compromising sensation and blood flow.
- Offer different size gloves (since workers' hands vary) and keep them in good condition.
- Replace greasy, dirty, or worn out gloves.

Vibration

Localized vibration from hand-held power tools can lead to decreased sensation in the hands and increase the force required to grip. This may lead to hand-arm vibration syndrome.

Localized Vibration

- Balance and maintain tools to decrease vibration and grip force.
- Decrease use of vibrating tools.
- Use speed adjustments to reduce vibration.
- Ask vendors for tool vibration characteristics, appropriateness for task, allowable length of exposure time, and maintenance schedules.

- Reduce vibration by using cushions, air-cushioned cylinders, air shutoff clutches, or isolation mounts.
- Change the process to reduce use of vibrating tools.
- Use low-vibration tools when possible. Are manufacturer accessories, substitutes, or revisions available that might decrease vibration?
- Keep tools in good working condition. Keep chisels and cutters sharp, replace shock absorbers and bent shafts, and maintain internal tool workings, such as pneumatic cylinder stops, and external devices, such as grips and handles.
- Decrease vibration transmission with tool stands, isolated fixtures or handles, and vibration absorbing material.
- Break up the task to avoid constant vibration exposure.
- Use supports to rest tools.
- Keep hands warm. Cold can lead to hand-arm vibration syndrome attacks through reduced circulation.

Whole-body Vibration (driving on uneven surfaces, heavy equipment vibration, etc.)

- Decrease uneven surfaces.
- Maintain equipment.
- Reduce speed.
- Use dock levelers, proper ramps, and adequate vehicle suspension to minimize travel vibration.
- Consider antifatigue mats and a sit/stand or lean seat for standing workers.
- Add rest breaks to avoid constant vibration exposure.

- Sitting at angles of 90 degrees (back to thigh) or less should be discouraged. Use lumbar supports.

Work Organization

Work pace and duration, overtime, unfamiliar work, lack of task variation, and recovery time are all work organization issues that contribute to CTDs. Use more than one strategy when several risk factors are present and alternate physical with mentally demanding tasks, and long-cycle with shorter cycle tasks.

Reduce Bending

- Use tilters, lift tables, work dispensers, and other mechanical assists.
- Raise the work to a better height or lower the employee.
- Provide and keep all material at waist height.

Decrease Twisting

- Provide all materials and tools in front of the employee or use adjustable swivel chairs.
- Use tilters, conveyors, chutes, slides, or turntables to bring work to the worker.
- Provide sufficient work space for the whole body to turn. Remove obstacles.

Revise Reaching

- Decrease horizontal reaches over 16 inches by moving tools, materials, and controls closer.

- Decrease size of materials or containers.

Lessen Lifting and Lowering

- Use mechanical aids: slides, chutes, lift tables, lift trucks, cranes, hoists, balancers, gravity dumpers, work dispensers, and conveyors.
- Raise the work or lower the worker.

Reduce Reach

- Provide grips, handles, or better access.
- Improve work layout.

Decrease Pushing and Pulling

- Use conveyors, powered trucks, slides, and chutes.
- Reduce force with lighter loads, air bearings, ball caster tables, four-wheeled hand trucks, dollies, or carts and pullers.
- Use friction to your advantage. Reduce it when objects are slid. Increase it for holding tasks.

Mimimize Carrying

- Decrease carrying by pushing or pulling. Use carts, conveyors, air bearings, ball caster tables, monorails, slides, chutes, lift trucks, hand trucks, and dollies.
- Lower weight by decreasing size, capacity, number of objects, weight, and load of containers.

- Reduce twisting motion.
- Provide an adequate work area.
- Place work in front of the worker.
- Put objects used the most closest to the worker.
- Use carts, turntables, containers, chutes, slides, swivels, and clamps.
- Provide adjustable swivel seats for those who sit and turn.
- Reduce reaching, especially for smaller workers, to avoid awkward positions.
- Place tools and controls within 16 inches of the worker.
- Reduce the size of objects being moved.
- Allow workers to move around objects.
- Put heavy objects as close to the worker as possible. Decrease the number of objects lifted.
- Reduce bending. Use mechanical aides such as scissor lifts, lift tables and dispensers.
- Place work at an appropriate level.
- Keep material at waist height.
- Reduce lifting and lowering.
- Use lift trucks, trollies, cranes, containers, elevators, lift tables.
- Decrease the weight and or capacity of containers.
- Reduce the number of objects being lifted.
- Be cautious about how pallets are lifted. Reduce the horizontal reach.
- Change the shape of an object, provide handles or grips to improve access to the object being lifted.
- Reduce pulling and pushing.
- Reduce friction by using rollers and conveyors.

Figure 6.2 Ergonomics Checklist

- Reduce repetition.
- Increase the number of muscles used.
- Use multifunction tools.
- Allow rest breaks. (A break should be one minute every 8 to 9 minutes or 5 to 10 minutes every hour for data entry or keyboard work.)
- Increase the number of different tasks done by increasing work cycle time. Work enlargement allows each worker to do many tasks, not just one.
- Reduce the force required to lift by changing shape or size of objects.
- Balance tools.
- Use air shutoff or external bars to control force.
- Use handles that are easy to grip.
- Wear proper clothing. Avoid anything too tight or too loose.
- Wear gloves to protect palms.
- Smooth or round out any sharp edges.
- Revise posture. Place work so that elbows can remain at the sides and at 90 degrees of flexion to avoid forearm rotation.
- Minimize standing. Cushion the floors and wear cushioned shoes.
- Alternate sitting and standing. Consider a sit/stand chair. Avoid using foot controls when standing.
- Use two hands instead of one.
- Use a power grip rather than a pinch grip.
- Grasp objects at their center of gravity.
- Provide a cushion for the hands during lifting tasks.
- Avoid awkward positions.
- Provide a proper chair, lumbar support, and appropriate level of work surface for sitting.

Figure 6.2 (cont'd) Ergonomics Checklist

Computer or Visual Display Terminals (VDT)

- Reduce glare.
- Place at a proper angle to eyes—3 to 4 inches below eye level, lower for bifocal wearers.

Filters are controversial. One study found they did not reduce eye strain or fatigue.[8]

Ergonomics Checklists

Ergonomics checklists are often helpful to categorize problems and organize interventions.

Figure 6.2 is based on the proposed OSHA ergonomics standards.

Putting It Together

- Fitting work to the worker is more than following a checklist, although the ones offered here and elsewhere can help. Fitting work to the worker requires at a minimum:
    1. Measuring the height of the work surface.
    2. Measuring the length of reach.
    3. Measuring fit or clearance.
- Work enlargement can be one of the most successful strategies to improve worker satisfaction. It includes:
    1. Decreasing repetitiveness of the work tasks.
    2. Enhancing job rotation.
- In the long run, fitting the work to the worker, instead of the reverse, is the most effective method to reduce the

incidence, cost, and liability of cumulative trauma disorders.

- Remember that ergonomics deals with more than musculoskeletal conditions. It includes redesigning the workplace, including tools, equipment, and work stations, so workers can perform at their best.

Endnotes

1. Morse, L.H., and L.J. Hinds. Women and ergonomics. *Occupational Medicine: State of the Art Reviews* 8(4):721-732, 1993.
2. Sommerich, C.M., McGothlin, J.D., and W.S. Marras. Occupational risk factors associated with soft tissue disorders of the shoulder: A review of recent investigations in the literature. *Ergonomics*, 36(6):697-717, June 1993.
3. Silverstein, B.A., Fine, L.J., and T.J. Armstrong. Carpal tunnel syndrome: Causes and a prevention strategy. *Semin Occupational Medicine* 1:213-221, 1986.
4. Eastman Kodak Company, *Ergonomic Design for People at Work*, Vol. 2. New York: Van Nostrand Reinhold, 1986.
5. Armstrong, T.J. *An Ergonomics Guide to Carpal Tunnel Syndrome*. Cincinnati: American Industrial Hygiene Association, 1983.
6. CFR1910. OSHA Proposed Ergonomics Standard, 1995.
7. See note 6.
8. Scullical, L., Rechichi, C., and C.A. De Moja. Protective filters in the prevention of asthenopia at a video display terminal. *Perceptual and Motor Skills* 80(1):299-303, February 1995.

7

Diagnosis, Prevention, and Treatment of Common CTDs

"Physicians get paid handsomely for entertaining their patients while they get better on their own."
—*Sir William Osler, M.D.*

Chapter Objectives

After completing this chapter, the reader will:
- Understand the conditions and diagnoses of common cumulative trauma disorders.
- Be able to recognize the early symptoms of cumulative trauma disorders.
- Understand muscle retraining and why early return to function after an injury is important.
- Know the basics of current CTD treatment recommendations and guidelines.
- Describe how to prevent reinjury and assist injury recovery.
- Know how to develop an effective modified duty program.

Case Study

Not Everything Is a CTD

A 61-year-old nurse with a numb right arm had surgery for carpal tunnel, but did not improve. A subsequent CAT scan revealed a herniated cervical disc. After neck surgery, her numbness finally resolved.

Diagnosis and Treatment Options

This chapter provides practical guidelines for CTDs, including prevention, diagnosis, treatment, rehabilitation, and return-to-work issues. A review of work activities and the workplace is critical. CTDs can often be prevented by creating safer, more comfortable work stations, by rotating tasks to break the cycle of repetition, and by simple stretches. For example, a typist can be instructed to perform wrist stretches, shoulder rolls, and neck rotations every hour. A worksite visit, demonstration of job tasks, and a team approach, involving safety staff, medical practitioner, employee health nurse, ergonomist, management, and worker, can prevent and ameliorate many occupational CTD cases. For acute care, good results can be obtained by well-organized treatment guidelines. During rehabilitation, identify local programs that have proven to be effective and choose interventions that are cost-effective.

Common CTD conditions are listed in Table 7.1. Note that CTD is used for classification purposes; for the individual patient, a more specific diagnosis such as carpal tunnel syndrome is preferable. Also the association of the following with cumulative trauma is subject to controversy (see discussion in previous chapters).

Table 7.1 Conditions That May Be Associated with CTDs

Neck	Tension neck syndrome or tension myalgia Cervical strain
Arm	Epicondylitis de Quervain's tenosynovitis or syndrome Tendinitis, bursitis, and tenosynovitis Carpal tunnel syndrome Ulnar tunnel syndrome Thoracic outlet syndrome Pronator syndrome Radial tunnel syndrome Posterior interosseous syndrome Anterior interosseous syndrome Cubital tunnel syndrome
Shoulder	Bicipital tendinitis Acromioclavicular syndrome Frozen shoulder or adhesive capsulitis
Back	Lumbar sacral syndrome

Diagnoses and Treatment of Specific CTDs

Tension Neck Syndrome or Tension Myalgia

Symptoms include an achy stiff neck, at times associated with a headache and spasm of the neck and upper back muscles (trapezius muscle). It occurs among dental hygienists, cashiers, small parts assembly workers, key punch operators, word processors, typists, packers, and others who contract the neck and upper back muscles. Inquire about symptoms that travel or radiate from the neck to the arms or elsewhere and any associated numbness, tingling, or burning sensation, as these may indicate a more serious injury, such as a herniated disc in the cervical spine.

Treatment is conservative and includes therapeutic message gentle range of motion and strengthening exercises, relaxation techniques, and at times a soft cervical collar. An individual should be made aware of posture, keeping the chin in rather than bending the neck forward. Stretching exercises can be helpful. These include bringing the chin to the chest and holding it; extending or bending the neck back looking upward; turning the head to the side and touching the chin to the right and then the left shoulder; and then placing the ear on the right and then the left shoulder. Each of these maneuvers should be held for five seconds at a time and may be done in the shower after allowing hot water to warm the neck.

Cervical Strain

Also known as cervical syndrome, cervical strain is similar to tension myalgia of the neck. Inquire about radiating discomfort down either or both upper extremities, decreased strength, numbness, and tingling and burning sensations in either or both upper extremities, as this may indicate more serious pathology, such as a herniated disk or cervical stenosis caused by bony narrowing of the spinal canal, which can occur with age or injury. Range of motion of the neck may be restricted by pain. This syndrome is encountered in workers who flex and extend their necks and who work in awkward positions, including clerical workers, dental personnel, painters, data entry workers, and cash register operators.

Conservative treatment is usually indicated, including therapeutic massage, physical therapy, gentle range of motion and conditioning exercises, as with tension neck syndrome, and a soft cervical collar if necessary for a few days (prolonged use is discouraged to avoid neck muscle weakness).

Epicondylitis

Lateral epicondylitis is commonly known as tennis elbow, and medial epicondylitis is often called golfer's elbow. Both conditions occur from repeated forced rotation and torquing of the forearm. Pain is noted on palpation of the epicondyles. The test for lateral epicondylitis involves extending the elbow and resisted extension of the wrist. Pain occurs at the extensor muscle group at the elbow and the lateral epicondyle. The test for medial epicondylitis is performed with the elbow flexed to 90 degrees and resisted wrist flexion. Pain occurs with palpation at the medial epicondyle.

These conditions are most commonly encountered where the elbow is repeatedly moved during the work process, as with musicians and construction, clerical, and assembly workers. Conservative measures are indicated, including physical therapy, range of motion exercises, and counter-pressure straps. Occasionally, localized injections are advised, and sometimes surgery is required. Workers are advised to keep the elbow at 90 degrees of flexion.

Tendinitis, Bursitis, and Tenosynovitis

Tendons are covered by a synovial sheath (the shoulder is an exception). High areas of stress and leverage are cushioned by bursa. Inflammation can occur in both structures. Direct trauma is not always present or recalled, and occupational factors are not always the cause, particularly in older individuals.[1]

Tendinitis may occur in the upper and lower extremities and is encountered when the tendon-muscle connection becomes stressed or weakened. The tendon sheath may become inflamed, a condition called tenosynovitis, and develop swelling or an effusion. At times the tendon may calcify, particularly in the shoulder. If the tendon sheath is thickened it will cause the tendon not to slip or move as easily, which leads to stenosing tenosynovitis (see de Quervain's syndrome and trigger finger, or stenosing tenosynovitis crepitans). Repetitive trauma can be a precipitating cause of trigger finger, particularly when hand tools are used among buffers, machine tool operators, assemblers, grinders, packers, and sewer workers. Treatment is conservative, including splinting, localized corticosteroid injections and surgery if no improvement occurs.

Rotator Cuff Tendinitis or Supraspinatus Tendinitis

This condition involves the tendon of the supraspinatus muscle, which abducts the humerus under the acromion. In a shoulder abducted position it will slide against the acromion, causing pain and inflammation. It is common in workers who keep the shoulder abducted with the elbow extended, including painters, riveters, construction workers, and welders. Typically, there is pain between 70 and 100 degrees of shoulder abduction. The drop-arm test involves holding the shoulder at 90 degrees of abduction and pressing down on the arm. The test is positive if the individual is unable to keep the arm up.

The impingement test for rotator cuff tendinitis involves holding the shoulder at 90 degrees of abduction and then internally and externally rotating it. Discomfort during the maneuver indicates a positive test.

Treatment is usually conservative, including physical therapy, instruction to externally rotate the humerus prior to reaching overhead, range of motion exercises to prevent a frozen shoulder, heat, anti-inflammatory medications, and localized corticosteroid injection. Surgery may be indicated if no improvement occurs.

Carpal Tunnel Syndrome (CTS)

The incidence of CTS is estimated at 0.5 to 3.0 percent among adults.[2-5] CTS is the most common compression syndrome of a peripheral nerve.[6] It is caused by the median nerve being compressed as it travels through the carpal tunnel of the wrist. Symptoms include numbness and tingling in the median nerve distribution of the hand, including the index and middle fingers, the inside of the ring fingers (radial aspect), the palm side of the thumb, and the dorsal aspect of the index, middle and ring fingers (radial side). If untreated, atrophy of the thenar muscles can occur.

A classic CTS symptom is awakening with numbness and tingling in the hand. It is relieved by shaking or flicking the hands.

Carpal tunnel syndrome in the workplace has been attributed to compression of the median nerve at the wrist.[7-12] Recent studies suggest that edema, caused by pressure or vibration of the distal small nerves and glabrous skin receptors is contributory.[13-15] This helps explain negative electrical studies in certain carpal tunnel syndrome cases.[16]

CTS is most common among clerical workers, word processor operators, assembly workers, grinders, packers, brick layers, cashiers, and others who undergo repetitive forceful wrist motion.[17] There is considerable controversy over the work-relatedness of CTS.[18,19] Some factors, such as aging, body mass index, and wrist rations, may lead to nerve slowing at the wrist to a greater degree than work activities.[20]

CTS also may be caused by other conditions such as gout, rheumatoid arthritis, familial recurrent polyneuropathy, and pregnancy. It is more common among woman, perhaps due to their smaller carpal tunnel and other factors, such as fluid retention during menses.[21]

Figure 7.1 outlines CTS prevention measures.

Situations to Avoid
1. Repetitive wrist flexion, extension ulnar or radial deviation (side to side movements).
2. Prolonged, firm repetitive gripping or pinching, especially when the wrist is deviated from a neutral position.
3. Direct or repetitive pressure over the palm and wrist area.
4. Gripping with the upper extremities elevated.
5. Hands in one position for an extended period, especially when gripping or pinching.
6. Dragging, jerking or tugging movements.
7. Clothing, watches, and jewelry that restrict the wrist.
8. Using the palms to pound.
Techniques that Prevent CTD
1. When driving, keep a relaxed grip on the steering wheel. Straighten the fingers, one hand at a time, frequently.
2. Position the hands at no higher than three and nine o'clock on the steering wheel, especially since airbags might strike hands held at the formerly recommended two and ten o'clock positions.
3. Relax fingers when pinching or gripping. A tight grip increases muscle-tendon unit pressure within the carpal tunnel.
4. When sitting, working or lying, keep wrists in a neutral position or in slight extension (can support by using a wrist bar, a pillow, or the arm or leg of a chair.
5. During sleep, wear a wrist splint to avoid wrist flexion, if recommended by a health care provider.
6. Lift with both hands instead of one hand, if possible.

Figure 7.1 Carpal Tunnel Syndrome Prevention Measures

Diagnosis of CTS

Tinel's sign (tapping over the median nerve at the wrist) and Phalen's sign (holding the wrists in hyperflexion for up to one minute) are suggestive of CTS if numbness, tingling, or pain occur in the hand or fingers. Gilliat's test (inflating a blood pressure cuff for one minute) and McMurthey's test (local pressure in the palmar surface over the carpal tunnel) lead to pain in the median nerve distribution if CTS is present.[22] Loss of thumb abduction and opposition strength may occur. See Table 7.2.

Treatment

Conservative treatment is usually indicated, including *neutral splints* (avoid splints that cock up or place the wrist in excessive extension) nonsteroidal anti-inflammatory drugs, and ergonomic modifications, such as a keyboard wrist bar to maintain the wrist in a neutral position. Steroid injections are occasionally useful. Carpal tunnel release either by open or endoscopic methods has a success rate of 70 percent to 95 percent.[23]

Post surgery symptoms may recur or persist and Tinel's sign and Phalen's test can remain positive after two years.[24] Newer endoscopic carpal tunnel surgery has promise. There is less tissue disruption during the procedure compared with the more traditional approach, but complications are a concern, including postoperative scar formation, inflammation and possible laceration of the superficial palmar arterial arch. Of 117 procedures reviewed by Australian researchers, 73 were rated as having excellent results. The remainder had good to poor results.[25] Another study compared traditional versus endoscopic surgery in individuals who had both procedures and found no post surgery differences, though endoscopy was preferred because of earlier return to activities.[26] A nonsurgical laser treatment has also demonstrated some benefit.[27]

Table 7.2 Diagnostic, History, Physical Examination Studies and Tests for Carpal Tunnel Syndrome

Magnetic resonance imaging (MRI)	Found a causative lesion in 32 out of 72 hands. Findings were confirmed in 16 of 24 hands during surgery.[28]
Nerve conduction tests	Testing thick nerves and thin fibers may provide further information. CTS nerve dysfunction is largely found in thick myelinated fibers, though certain thin nerve dysfunction was also found during thermotesting.[29] Vibrometer testing for CTS has some promise,[30] but can be compounded by variables such as menses-related fluid retention, which varies in women, and which can increase vibration thresholds.[31]
Examination	Phalen's test and Tinel's sign; grip and pinch strength.
Interview	Numbness, weakness, nocturnal symptoms, pain, and swelling.[32]

Ulnar Tunnel Syndrome

This syndrome involves the ulnar distribution of the hand. It may occur among those who bump or press the inside of the elbow against a work surface or those who compress the tunnel of Guyon. Typically, there will be numbness, tingling on the outside or ulnar aspect of the little finger. Tinel's sign can be tested over the ulnar nerve at Guyon's tunnel of the wrist. A related condition is cubital tunnel syndrome, which also affects the ulnar nerve more proximally at the medial epicondyle of the elbow. Symptoms are similar to hitting your "funny bone." Treatment mimics carpal tunnel syndrome previously described.

It is important to delineate whether hand and wrist complaints are due to median or ulnar nerve entrapment, or to tendinitis, which is characterized by synovial thickening and pain on resisted wrist maneuvers. In contrast to carpal tunnel syndrome, nocturnal symptoms are infrequent in tendinitis.

Upper Extremity Nerve Entrapment Syndromes

Thoracic outlet syndrome, bicipital tendinitis, frozen shoulder syndrome or adhesive capsulitis and acromioclavicular syndrome involve the upper extremity. See Figure 7.2 for risk factors.

Usually bilateral (two sides), but can be unilateral (one side)
• smoking
• diabetes mellitus
• alcohol
• heredity (familial recurrent polyneuropathy)
• acromegaly
• hypothyroidism
• rheumatoid arthritis
• amyloidosis
• pregnancy
• menses
• malnutrition
May be unilateral, but can be bilateral
• gout
• trauma
• carpal tunnel syndrome
• ulnar tunnel syndrome
• cubital tunnel syndrome
• radial tunnel syndrome (sometimes associated with lateral epicondylitis or tennis elbow)
• infection
• thoracic outlet syndrome[33]

Figure 7.2 Risk Factors for Nerve Entrapment

Thoracic Outlet Syndrome (TOS)

TOS is caused when the nerves of the brachial plexus and brachial artery branches are pressed between the muscles of the shoulder and the neck. Causes include improper posture and less commonly neurogenic or vascular disorders. Symptoms include numbness or tingling and pain in the affected upper extremity, especially when the shoulders are pulled back and the arm is raised. Headaches may be associated with TOS. The presence of a decreased pulse during Adson's test is suggestive. To perform Adson's test, externally rotate the shoulder with the chin thrust forward. Diagnostic tools include electromyographic (neurogenic TOS) and Doppler studies, somatosensory response, and electromyography-guided scalene blocks. Wright's test and the military brace test may also be positive. Radiographs may show a cervical rib causing TOS, or hardening of the brachial artery or unusual muscle placement and insertions.

TOS sometimes occurs in occupations where there is frequent reaching and lifting above the shoulder level and repetitive use of the shoulder girdle with fine motor work such as data entry, grocery checking, auto repair, luggage manufacturing, construction, stockroom, and assembling, as well as in individuals who encounter pressure over the shoulder girdle, such as rangers who wear backpacks and letter carriers who carry mail packs.

Treatment is usually conservative, and TOS may respond to weight loss, physical therapy, and strengthening of the shoulder girdle, especially with elevation exercises.[34] If no improvement results with conservative treatment, surgery may be beneficial for some indications.

Bicipital Tendinitis

This condition is often associated with supraspinatus tendinitis. It involves pain in the glenohumeral joint and over the bicipital tendon, which is usually tender. Flexion of the biceps and forearm supination against resistance are often uncomfortable. This condition occurs in individuals who work and reach overhead, including construction workers, stockroom personnel, shipping clerks, assembly workers, and window cleaners. Treatment is the same as for supraspinatus tendinitis.

Acromioclavicular Syndrome

Pain over the acromial joint is characteristic. A confirmatory test is to ask the worker to push downward with the hands against resistance. The examiner then taps on the clavicle, reproducing the pain. It often occurs when stress is placed on the shoulder joint by pushing on a work surface at waist level, including grinding, construction work and assembly. Treatment is conservative, including nonsteroidal anti-inflammatory drugs. Local injections of steroids may have some benefit, but sometimes surgical restriction of the distal clavicle is indicated.

Frozen Shoulder or Adhesive Capsulitis

This condition may result from contracture of the tissue around the glenohumeral joint and thickening of the tendons and bursa. Symptoms involve limitation of movement and severe pain with movement of the shoulder. It may result from immobilization after a shoulder injury. Early initiation of physical therapy and range of motion exercises are critical.

Pronator Syndrome

This syndrome is due to median nerve entrapment at the elbow. Carpal tunnel syndrome is median nerve entrapment at the wrist. Symptoms include pain in the volar, or palm side, of the forearm and numbness in the thumb, index, long finger, and radial half of the ring finger. Physical examination is similar to carpal tunnel syndrome. Pronation, long finger flexion, and elbow flexion against resistance are painful. Tinel's test is positive at the forearm, and pronation with elbow flexion is weak. Abnormal electrical studies are revealing, though rarely encountered.

Treatment includes rest, splint with the elbow in flexion and forearm in pronation, and surgery if no improvement results.

Radial Tunnel Syndrome

This syndrome involves entrapment of the radial nerve distal to the elbow. Symptoms include pain over the extensor forearm when making a fist, or on forearm supination or wrist dorsiflexion. Abnormalities on examination include weakness with wrist, thumb, and finger extension. Confirmatory tests include resisted long finger extension.

Treatment involves rest, a splint with elbow in flexion, forearm in supination, and wrist in extension, or surgery if no improvement occurs.

Posterior Interosseous Syndrome

This condition also involves compression of the radial nerve at the forearm with tenderness four finger breadths distal to the elbow at the Arcade of Frohse. Pain in the proximal forearm and finger and thumb and partial wrist extension weakness (can extend the wrist, but it deviates radially) can occur. Electrical studies are diagnostic, and the treatment is the same as for radial tunnel syndrome.

Anterior Interosseous Syndrome

This condition is due to median nerve compression distal to the elbow. Symptoms include pinch weakness and pain in the mid-volar forearm. There is weakness in index finger and thumb pinch (unable to make the "OK" sign), flexor pollicis longus, flexor digitorum of the index finger, pronator quadratus, and with pronation with elbow flexion. Electrical tests are helpful. Treatment includes rest along with splinting the elbow in flexion and the forearm in pronation.

Cubital Tunnel Syndrome

Cubital tunnel syndrome is due to ulnar nerve entrapment at the elbow. Some make a distinction between ulnar nerve compression at the ulnar groove (ulnar succus syndrome) and compression more distal as the ulnar nerve passes between the two heads of the flexor carpi ulnaris (cubital tunnel syndrome). Symptoms are pain in the medial elbow and numbness in the little finger and ulnar half of the ring finger. Signs include decreased sensation in the little finger and ulnar half of the ring finger, intrinsic muscle weakness, and positive Tinel's sign at the elbow. Phalen's test, Allen's test, palmaris brevis sign and Froment's paper sign are often positive.

Treatment includes rest, and elbow extension in a night splint. Surgery is an option if no improvement occurs.

de Quervain's Syndrome

First extensor tenosynovitis, or de Quervain's syndrome, occurs in the first dorsal extensor compartment of the wrist, which extends from the thumb to the wrist. The disease is common in workers who repetitively deviate their wrist side to side in an ulnar and radial direction. Some of these occupations are cashiers, mechanics, shippers, stockers, carpenters, and butchers, and especially in clerks who write on forms with multiple copies and have to write firmly. A congenital anomaly in which the extensor pollicis brevis tendon has its own compartment is another cause of de Quervain's.

The first extensor compartment is tender and the Finkelstein test is positive. In this test, the examiner wraps his fingers over the patient's thumb and ulnar deviates the wrist. Pain occurs at the radial styloid and along the first tendon extensor sheath.

Treatment includes a splint, called a thumb spica splint, that incorporates the thumb. Corticosteroid injections into the first extensor sheath are sometimes beneficial. Surgery is often indicated.

Vibration Syndromes

The use of vibrating tools can lead to an array of symptoms in the upper extremities, including pain, numbness, and dysfunction. Vibration-induced white finger resembles Raynaud's phenomenon. Hand-arm vibration syndrome may mimic cervical or peripheral nerve entrapment of the ulnar or median nerves.[35,36] Treatment is similar to that of Raynaud's phenomenon, and involves removal from vibration, and the use of splints, nonsteroidal anti-inflammatory drugs, calcium-channel blockers. Surgical release of entrapment may be indicated in unresponsive cases.

Lumbar Sacral Syndrome

(The following sections of this chapter are adapted from Peate, WF, Occupational musculoskeletal diseases. *Primary Care* 21:313-327, 1994, with acknowledgment to W.B. Saunders Co.)

Low back injuries are the most common and expensive of CTDs, and treatment of low back pain in working adults is more costly than any other disease entity.[37] So important is back pain that the federal government has recently released guidelines for the treatment of low back pain known as Clinical Practice Guidelines Number 14, Acute Low Back Problems in Adults: Assessment and Treatment.[38] It is published by the Agency for Health Care Policy and Research (AHCPR). These guidelines are outlined in Figure 7.3.

Beneficial
Over-the-counter medication (acetaminophen, nonsteroidal anti-inflammatory drugs)
Exercise (conditioning, walking, swimming)
Spinal manipulation (in the first month of symptoms)
Unsupported by research
TENS (transcutaneous electrical nerve stimulation)
Acupuncture
Spinal traction
Biofeedback
Lumbar supporters, belts, corsets
Heat, massage, ultrasound, electric stimulation
Extended bed rest (more than four days)
Epidural steroid injections
Oral steroids

Figure 7.3. AHCPR Guidelines for Treating Uncomplicated Back Pain

Social Factors

Nonorganic factors have an important role in the management of occupational CTDs. A poor job performance appraisal,[39] smoking, inadequate sleep, and substance abuse have all been implicated as contributors to occupational musculoskeletal disorders.

The expenses and recovery times for work injuries may be double those for non-work-related injuries, according to the Minnesota Workers' Compensation study, especially for low back pain, which is difficult to verify because objective findings are present in only 50 percent of cases.[40] In contrast, for upper extremity fractures, in which objective findings are evident, there was no significant difference in recovery times between work-related and non-work-related injuries. See Table 7.3.

Diagnostic Uncertainty

Low back problems and other CTDs often lack a firm diagnosis or a definite causation. As many as 85 percent of low back pain cases cannot be firmly diagnosed.[41] The worker may be unable to recall a precipitating event because direct trauma is rarely a cause of CTD's. Most are the accumulation of twisting, repetition, overloading, and prolonged poor posture.

Specific and Active Treatment Plan

Diagnostic uncertainty should not lead to indecision. The injured worker should be provided with a specific rehabilitation program that is active and initiated early. With this approach success has been confirmed in several low back pain studies. Donelson et al found a 98 percent improvement in low back pain if an exercise program was begun in the first four weeks after injury and only 80 percent improvement if the program began after that time period.[42]

Table 7.3 Comparison of Work-Related and Non-Work-Related Injuries[43]

Diagnosis	Cost ($)	Duration of Treatment (days)
Low back pain		
Work-related	308	21
Non-work-related	132	10
Upper extremity fracture		
Work-related	224	23
Non-work-related	226	30

The Quebec Task Force on Spinal Disorders determined that active modalities, such as exercise and return to activities of daily living, were more effective than passive behaviors, such as bed rest, repeated hot packs, diathermy, and massage.[44] Cady established that workers who maintained good fitness recovered quicker after an episode of acute low back pain and had less risk of chronic pain than those who where not fit.[45] Physical therapy is appropriate for acute symptom relief and for teaching exercises and proper body mechanics. The number of sessions should be specified, the worker should be offered a firm follow-up date to monitor progress, and a recall effort should be made if he or she fails to keep the appointment.

Value of a Graduated Return to Activities of Daily Living and Work

Unfortunately, some health care providers rely on the absence of pain, rather than the restoration of function, as a measure of recovery. When an injured worker who is finally pain free but now physically deconditioned is released to work, he or she is susceptible to reinjury. Mayer et al found that pain was not a contraindication for a progressive return to productive activity.[46] Eighty percent of the chronically disabled return to work with residual low back pain. Waddell described the hazards of delayed return to the activities of daily living and work. Only 50 percent of workers with low back pain who are off work for six months ever return to work. Of those not working for two years, only 2 percent ever work again.[47] Providers should strive to enable, not disable.

Fear of Impairment

Clear and appropriate communication during the initial visit for low back pain is essential. The injured worker should be reassured that most CTDs are temporary. Fifty-five percent of patients with acute low back pain believe their condition will be chronic and disabling,[48] even though fewer than 5 percent of those with low back pain remain off work more that six to twelve months after their injury.[49,50]

A cumulative trauma disorder should be described to the injured worker as a temporary condition, uncomfortable, but brief. Permanent rules to live by should be offered. For example, to avoid recurrence, someone with a history of low back pain should avoid bending at the waist.

The injured worker should be prepared for a roller coaster recovery. Symptoms may worsen for no apparent reason, and then improve.

Radiograph (X-ray) Expectations

Many patients will expect an x-ray, even for minor injuries. It should be explained that such studies are not without risk, since radiation can cause leukemia and have other effects, and that they offer little information in most back injuries or CTDs. The worker should be offered a recovery timetable. One might say if no improvement is noted by the next visit in three to four weeks, or as indicated, further studies might be considered.[51]

Symptoms

Ask about the onset, nature, and duration of symptoms, and about relieving or aggravating factors. The worker should rate the pain on a 10-point pain scale, with 0 equal to no pain and 10 equal to the worst pain.

Past History

The lifetime incidence for low back pain is 70 percent,[52] and many injured workers treat themselves or consult alternative medicine providers. Ask the worker about having used ice or heat at home, having worn lumbar belts (check that these are worn correctly over the lumbar area), having taken over-the-counter and topical medications, and having used inversion therapy. Because the head is upside down in inversion therapy, this modality is contraindicated in hypertension and glaucoma.

Prior low back pain is associated with a longer recovery in those who have had three or more previous episodes.[53]

Review of Symptoms

Three percent of low back pain is due to non-musculoskeletal causes.[54] Table 7.4 summarizes common symptoms and related conditions.

Physical Examination

Observe the individual's gait. Those with moderate-to-severe spinal stenosis gradually flex forward—the stoop test. The presence of foot drop may indicate a herniated nucleus pulposus at L4-5, and those who prefer to stand rather than sit may also have a herniated nucleus pulposus. Those who would rather sit than stand may have spinal stenosis. Check for an uneven wear pattern on shoes. This could indicate possible lower extremity or hip disorder, which may contribute to low back pain. Palpate the skin for roll tenderness, trigger points (fibromyalgia), and spasm.

Circumference

More than a one-inch or 2.2 cm difference[55] between the circumferences of the lower extremities when measured one breadth of the hand above or below the patella is significant for muscle wasting.

Range of Motion

Internal and external hip range of motion should be checked to rule out hip pathology in that joint.

Table 7.4 Common Symptoms and Related Conditions in Low Back Pain

Symptom	Condition
Pain at rest (herniated nucleus pulposus/osteomyelitis improves with rest), failure to improve after one month, unexplained weight loss, prior neoplasm, age more than 50 years.[56]	Neoplasm
Fever	Osteomyelitis
Buttock pain with sexual activity.	Pyriformis muscle injury
Dizzy and diaphoretic, low back pain.	Abdominal aortic aneurysm
Pain worse in the AM, improves with exercise, slow onset, duration more than 3 months.	Ankylosing spondylitis (usually begins before age 40)
Pain increases with walking, standing, or lumbar extension (better with stairs or stooping forward).	Lateral stenosis (Pseudoclaudication)
Radiating pain, pain with sitting, lumbar flexion and Valsalva's maneuver. Better with bed rest.	Herniated nucleus pulposus (Slipped or ruptured disc)
Pain with Valsalva's maneuver, sensory deficit in buttocks, perianal region and both legs. Change in bowel, bladder or sexual function.	Cauda equina syndrome (surgical emergency)
More pain with extension than with flexion.	Facet syndrome (often caused by flexion injury)[57]

Reflexes

The majority of cases of herniated nucleus pulposus occur between L3 and S1. Evaluate knee jerks (L3-4), posterior tibial reflexes present in 40 to 50 percent of normal individuals (L4-5), and ankle jerks (L5-S1).

Pulse rate and strength should be determined. Pulses are normal in pseudoclaudication (spinal stenosis), and decreased in claudication (hardening of the arteries).

Special Tests

In herniated nucleus pulposus (HNP), a positive straight leg raise test is present in 95 percent of cases, and a positive cross straight leg raise test has high specificity.[58] Conduct a confirmatory test. With the leg at the limit of straight leg raising tolerance, dorsiflex the ankle. Worsening of leg pain is suggestive of HNP. Patrick's test will distinguish hip and sacroiliac disease from lumbar disk problems. The lateral ankle is placed on the opposite knee and the hip externally rotated and abducted. A positive femoral stretch test suggests upper lumbar disc disease (L2-4). With the worker in a prone position, the knee should be flexed with the hip extended.

Waddell's signs are a means of measuring nonorganic low back pain (pain without a physical basis), a diagnosis that should be considered if three of five of the following are positive: 1) complaints of low back pain during simulation of spine loading (pressing downward on the head) or rotation turning the shoulders and hips in one piece; 2) nonorganic tenderness; 3) discrepancy between supine and sitting straight leg raise testing; 4) inappropriate sensory findings and 5) overreaction during the examination.[59]

Special Studies

In most cases of work-related low back pain, a complete history and physical examination are sufficient. Radiographs should be considered for the following conditions: significant trauma, past history of surgery or neoplasm (especially renal, lung, prostate, or breast), drug or alcohol abuse, steroid use or other evidence of immunocompromise, fever greater than 38 degrees Centigrade (101 degrees Fahrenheit), weight loss, rest pain, and neurologic findings, or if ankylosing spondylitis is suspected.[60]

Computerized axial tomography and magnetic resonance imaging can rule out a pathologic process or be used when surgery is considered, as with cauda equina syndrome. Be wary about false-positive results. Twenty to 30 percent of individuals with a negative history of low back pain will have a herniated nucleus pulposus on computerized axial tomography or magnetic resonance imaging.[61-63]

Guidelines for special studies are offered in Figure 7.4.

Laboratory Studies

The erythrocyte sedimentation rate is a nonspecific test to distinguish between inflammatory and noninflammatory causes of low back pain. HLA B27 is an assay to determine ankylosing spondylitis, but that diagnosis should be made on clinical grounds and sacroiliac films.

Treatment

For uncomplicated low back pain, that is with the absence of sciatica, neurologic deficits, or structural changes, two to three days of bed rest are sufficient.[64] Longer periods, those exceeding two weeks, can lead to permanent bone loss, especially in women.

- Magnetic resonance imaging, if available, is advised over computerized axial tomography for disc disorders.

- Computerized axial tomography is preferred for bony abnormalities.

- Magnetic resonance imaging is replacing myelography. Request magnetic resonance imaging with gadolinium for post surgery pain to distinguish between scarring and recurrent herniated nucleus pulposus.[65]

- A bone scan may be indicated to rule out cancer, occult fracture, or infection.

- Ultrasonography has a 70 percent accuracy rate for herniated nucleus pulposus versus 90 percent for magnetic resonance imaging.[66]

- Request electromyography nerve conduction studies if the above studies are negative or if a metabolic, hereditary, or a chemical-related disorder is a concern.

Figure 7.4 Guidelines for Special Studies for Low Back Pain

A Protocol

- Day 1, 2: Bed rest (firm mattress, bed board, or the floor) with bathroom and meal privileges. Avoid sitting up in bed to read or watch television. Begin easy painless stretches.
  Encourage self care and advise ice for acute symptoms (no more than 10 minutes every hour). For those who will not tolerate cold, offer

heat, 20 minutes every hour, three to four times a day. Warn not to fall asleep on a heating pad or burns may result. May alternate with ice.

- Day 3: Begin walking 20 minutes on a level surface every three hours, three to four times a day, and exercising to avoid deconditioning.

    Specific low back exercises are important as a recovery and injury prevention measure. They fall into two general categories. Avoid extension or MacKenzie exercises in spinal stenosis, spondylolysis, or facet syndrome. Flexion or Williams' exercises are contraindicated for posterior midline herniated nucleus pulposus. Exercise also induces a natural endorphin or pain-relieving effect.

- Day 4: Follow-up appointment with health care provider. Discuss modified duty, based on discussion with the patient's supervisor or human resources director. Instruct the individual to call if unable to tolerate the specific limitations prescribed (e.g., no lift, push, or pull greater than 20 pounds with frequent change of position) or if asked to exceed those restrictions at work. Lifting, twisting, and prolonged sitting restrictions may be needed for four to eight weeks.

    Ensure adequate lumbar support when sitting or driving. A rolled up towel is sufficient.

- Day 7: If the pain has not improved, or has worsened, or if sciatica (radiating leg pain) is present, consider further diagnostic studies.

Low Back Pain with Leg Pain

Sciatica treated conservatively will usually improve with the measures described here. Bed rest may be extended for a week. Surgery usually does not offer an improved recovery rate. Exceptions include injured workers with bowel or bladder dysfunction or progressive neurologic changes, such as foot drop.[67]

Medication

Nonsteroidal anti-inflammatory drugs, if tolerated, are often helpful. For sciatica, advise regular use rather than medication on an as needed basis. For patients with platelet aggregation concerns (bleeding disorders, etc.), a history of gastrointestinal blood loss or upset, consider a nonacetylated nonsteroidal anti-inflammatory drug, such as salsalate. If renal compromise is an issue, sulindac is suggested. Misoprostol, 100 mg four times daily, may be offered concurrent with nonsteroidal anti-inflammatory drugs for those with gastrointestinal intolerance, but it must be taken with food and avoided in pregnancy. The patient should be warned of diarrhea, which occurs initially in 20 percent of cases.

Narcotic agents for acute pain relief and muscle relaxants, such as cyclobenzaprine, have no place in long term therapy.

Other Treatment Modalities

Transcutaneous electrical nerve stimulation (TENS) and electroacupuncture for chronic low back pain are no better than placebo.[68] Traction is ineffectual, as 25 percent of body weight is required to reduce intradiscal pressures.[69]

Epidural injections are controversial,[70] and facet joint injections of steroids have been discounted by a well-controlled study.[71]

Spinal manipulation for the acute relief of low back pain is beneficial when limited in duration,[72,73] but there is variation in technique, monitoring may be difficult,[74,75] and it is contraindicated for herniated nucleus pulposus, spondylolysis, or spondylolisthesis.

Lumbosacral back supports are controversial and not a replacement for an adequate low back strengthening program. Offer instruction in their proper use and advise removal of the supporter periodically to prevent muscle deconditioning.

Treatment of Chronic Low Back Pain

Chronic low back pain is defined as a condition of more than seven weeks' duration, according to the Quebec Task Force on Spinal Disorders.[76] Further diagnostic tests are frequently indicated. For persistent cases, consider antidepressants, such as nortriptyline and trazodone, a physical medicine and rehabilitation evaluation, a pain management or psychological consultation, biofeedback, and work hardening.

Work hardening prepares the injured worker for reentry into employment through rehearsal in a simulated work setting. Under the guidance of a multidisciplinary staff of physician, psychologist, vocational rehabilitation specialist, and physical therapist return to work can be facilitated and the confidence of those with a chronic injury can be restored.[77] Rates for return to active employment range from 68 to 85 percent.[78-81] Unfortunately, there are few controlled studies, and considerable variation exists between programs. When possible, refer to a program approved by the Committee on Accreditation of Rehabilitation Facilities.

Prevention

Determine the worker's functional capacities: strength, range of motion, and endurance. Appropriate worker training in proper lifting is essential. In the past, workers were told to use their legs to lift while keeping their back straight because bending forward with rotation while carrying ten kilograms for example, increases lumbar disk pressure twenty fold.[82] However, if the load is too large to be brought between the legs, or if it is located horizontally away from the feet, then bending at the waist may be preferable. Intradiscal pressures are the same when the ratio of moment arm to load is constant.[83]

Instruct workers to hold heavy loads close to the body and to avoid sudden twisting of the waist. Workers should be reminded that low back pain occurs not only from lifting, but also from working on slippery and uneven surfaces and from push and pull activities.

Workers whose activities are physical should be informed that they are "industrial athletes," and should warmup before work and perform conditioning exercises at least three times a week to prevent injury.

Return-to-Work Issues

Figure 7.5 provides a form to complete for return to work authorization and restrictions.

Under the Americans with Disabilities Act, a medical examiner must demonstrate that a prospective employee's prior medical condition poses a direct threat to the safety and well-being of others, that no reasonable accommodation can be made, and that significant objective findings with proven predictive value are evident.

TO: _____
Mr./Ms. _____ is seeing you for:

☐ Second opinion
☐ Medical evaluation to complete an employment physical
☐ Evaluation and treatment
☐ Complete the work status form provided below.
  Modified duty is usually available—often in a position different than the employee's assigned department. You must include *specific restrictions*; "light duty is insufficient."

  Before placing an employee on *off-work status*, call our Occupational Health Network physician or his/her designee at _____
☐ Send a medical report with specific work status to: _____

RETURN TO WORK

Employee _____ Date of Injury _____
☐ Regular Duty Immediately   ☐ On ___/___/___
☐ Modified Duty—Available on ___/___/___ ☐ In a position different than the employee's assigned department.

SPECIFIC RESTRICTIONS

☐ No heavy lifting, heavy straining.	☐ Allow frequent change of position	☐ One-handed job only
☐ No heavy exertion or frequent bending	☐ Recommend work station evaluation	☐ No reaching above shoulder level
☐ Weight limit ___ lbs.	☐ No latex/powdered gloves	☐ No extensive walking or prolonged standing
☐ No stooping, crawling, climbing, or work on rough or oily surfaces	☐ Current workday capacity Can sit for ___ hrs. Can stand for ___ hrs. Can walk for ___ hrs.	☐ Patient can use hands for repetitive action such as: ☐ Simple grasping ☐ Pushing and pulling ☐ Fine manipulations
☐ Ground level work only	☐ Other_____	

PERMANENT RESTRICTIONS:_____

Signed_____ Today's Date_____
Name/Address (Please print) _____

Figure 7.5 Return to Work Form

Functional capacity evaluations, which analyze strength and range of motion and match the physical capabilities of the injured, recovered, or disabled worker with expected job duties may be of help in Americans with Disabilities Act accommodation issues.

Putting It Together

- Splints and supports are overused and their efficacy is undersupported. If used, they should be worn properly and chosen appropriately. A lumbar supporter, for example, should be worn over the low back and not the ribs, and a neutral splint should be chosen for carpal tunnel syndrome, rather than a cock-up splint.
- Always explore non-work-related activities as reasons for cumulative trauma disorders, such as home computer games, video games, crocheting, knitting, sewing, or gardening.
- Individuals with cumulative trauma disorders are often unable to recall a specific precipitating factor or an acute event leading to their disorder. Hence the name "cumulative." This factor leads to difficulties in determining whether a CTD is compensatory under many workers' compensation systems and to the disbelief among employers.
- Providers should 1) request and review job descriptions before returning an injured worker to work, and 2) provide specific guidelines. "Light duty" is not sufficient—be specific. An example: the worker may grip repetitively for up to 20-minute intervals separated by 10-minute rest breaks. The total repetitive gripping should not exceed two hours per day.

- Return to function and early consideration of modified duty is critical. The longer workers are away from work the less likely they are to ever return.[84]

- Work hardening can prepare injured workers for reentry to their former jobs by rehearsing work in a controlled and guided setting.

Endnotes

1. Meine, J. Pathogenesis of insertion tendinitis of the elbow in insurance medicine. *Zeitschrift fur Unfalchirurgie und Versichervrgs medizin* (Switzerland) 87(3):169-177, September 1994.
2. Moore, J.S. Carpal tunnel syndrome. *Occupational Medicine* 7:741-63, 1992.
3. Stevens, J.C., Sun, S., Beard, C.M., O'Fallon, W.M., and L.T. Kurland. Carpal tunnel syndrome in Rochester, Minnesota, 1961 to 1980. *Neurology* 38:134-138, January 1988.
4. Occupational disease surveillance: carpal tunnel syndrome. *Morbidity and Mortality Weekly Report*, 38:485-489, 1989.
5. Franklin, G.M., Haug, J., Heyer, N., Checkoway, H., and N. Peck. Occupational carpal tunnel syndrome in Washington State 1984-1988. *American Journal of Public Health* 81:741-746, June 1991.
6. Dawson, D.M., Hallet, M., and L.H. Millender. *Entrapment Neuropathies*, 2nd ed. Boston: Little Brown and Co., 1990.
7. Armstrong, T.J., and D.B. Carpal tunnel syndrome and selected personal attributes. *Journal of Occupational Medicine* 21:481-486, 1979.

8. Cannon, L.J., Bernacki, E.J., and S.D. Walker. Personal and occupational factors associated with the carpal tunnel syndrome. *Journal of Occupational Medicine* 23:255-258, 1981.
9. Feldman, R.G., Goldman, R., and W.M. Keyserling. Peripheral nerve entrapment syndromes and ergonomic factors. *American Journal of Industrial Medicine* 4:661-681, 1983.
10. Kaplan, P.E. Carpal tunnel syndrome in typists. *Journal of the American Medical Association* 250:821-20, 1983.
11. Katz, J.N., and M.H. Liang. Carpal tunnel syndrome and the workplace: Epidemiologic and management issues. *International Medicine* 9:66-73, 1988.
12. Masear, V.R., Hayes, J.M., and A.G. Hyde. An industrial cause of carpal tunnel syndrome. *Journal of Hand Surgery* American Volume, 11:222-227, 1986.
13. Gelberman, R.H., Szabo, R.M., Williamson, R.V., and M.P. Dinick. Sensibility testing in peripheral-nerve compression syndromes: An experimental study in humans. *Journal of Bone and Joint Surgery*. American Volume, 65:632-638, 1983.
14. Lundborg, G., Myers, R., and H. Powell. Nerve compression injury and increased endoneurial fluid pressure: A "miniature compartment syndrome." *Journal of Neurology, Neurosurgery and Psychiatry*, 46:1119-1124, December 1983.
15. Szabo, R.M., Gelberman, R.H., Williamson, R.V., Dollon, A.l., Yarn, N.C., and M.P. Dimick. Vibratory sensory testing in acute peripheral nerve compression. *Journal of Hand Surgery* American Volume, 9A:104-109, 1984.
16. Grund, A.B. Carpal tunnel decompression in spite of normal electromyography. *Journal of Hand Surgery* American Volume, 8:348-349, 1983.
17. Osorio, A.M., Ames, R.G., Jones, J., Castorina, J., Rempel, D., Estrin, W., and D. Thompson. Carpal tunnel syndrome among grocery store workers. *American Journal of Industrial Medicine* 25:229-245, February 1994.
18. Radecki, P. A gender specific wrist ratio and the likelihood of a median nerve abnormality at the carpal tunnel. *American Journal of Physical Medicine and Rehabilitation* 73:157-162, 1994.
19. Hadler, N.M. *Occupational Musculoskeletal Disorders*. New York: Review Press 1993:197.
20. Radecki, P. Variability in the median and ulnar nerve latencies: implications for diagnosing entrapment. *Journal of Occupational and Environmental Medicine* 37:1293-1299, 1995.
21. Dekel, S., and R. Coates. Primary carpal stenosis as a cause of "idiopathic" carpal tunnel syndrome. *Lancet*. Nov 10, 1979.
22. Buch-Jaeger, N., and G. Foucher. Correlation of clinical signs with nerve conduction tests in the diagnosis of carpal tunnel syndrome. *Journal of Hand Surgery* British Volume 19B, 1994.

23. Carpal Tunnel Syndrome. *Mayo Clinical Update*, Volume 9, Summer 1993.
24. Katz, J.N., Fossell, K.K., Simmons, B.P., Swartz, R.A., Fossel, A.H., and M.J. Koris. Symptoms, functional status and neuromuscular impairment following carpal tunnel release. *Journal of Hand Surgery* 20A (4) 549-555, July 1995.
25. Shinya, K., Lasnzatta, M., and W.B. Connolly. Risk and complications in endoscopic carpal tunnel release. *Journal of Hand Surgery*, Scotland, 70:222-227, 1995.
26. Futami, T. Surgery for bilateral carpal tunnel release compared in 10 patients. *Acta Orthopaedica Scandinavica*, Oslo, 66(2):153-155, 1995.
27. Weintraub, M.I. Laser Photobioactivation and neurolysis in carpal tunnel syndrome: A novel, nonsurgical approach. American Academy of Neurology 18th Annual Meeting Program, February 1996.
28. Seyfert, S., Boegner, F., Hamm, B., Kleindienst, A., and C. Klatt. The value of magnetic resonance imaging in carpal tunnel syndrome. *Journal of Neurology* 242:41-46, 1994.
29. Lang, E., Spitzer, A., Plannmuller, D., Claus, D., Handwerker, H.O., and B. Neundorfer. Function of thick and thin nerve fibers in carpal tunnel syndrome before and after surgical treatment. *Muscle and Nerve* 18:207-215, February 1995.
30. Gerr, F., Letz, R., Harris-Abbott, D., and L.C. Hopkins. Sensitivity and specificity of vibrometer for detection of carpal tunnel syndrome. *Journal of Occupational Environmental Medicine* 37:1108-1115, 1995.
31. White, K.M., Congleton, J.J., Huchungson, R.D., Koppa, R.J., and O.J. Pendelton. Vibromety testing for carpal tunnel syndrome: a longitudinal study of daily variations. *Archives of Physical Medicine and Rehabilitation* 75:25-28, January 1994.
32. Young, V.L., Seaton, M.K., Feely, C.A., Arfken, C., Edwards, D.F., Bawm, C.M., and S. Logan. Detecting cumulative trauma disorders in workers performing repetitive tasks. *American Journal of Industrial Medicine* 27:419-431, March 1995.
33. Bracker, M.D., Ralph, L.P. The numb arm and hand. *American Family Physician* 51:103-116, 1995.
34. Novak, C.B., Collins, E.D., and S.E. MacKinnon. Outcome following conservative management of thoracic outlet syndrome. *Journal of Hand Surgery* American Volume. 20A (4):542-548, July 1995.
35. Olson, N. Diagnostic tests in Raynaud's phenomena in workers exposed to vibration: A comparative study. *British Journal of Industrial Medicine* 45:426-430, 1988.
36. Pyykko, I. Clinical aspects of the hand-arm vibration syndrome: A review. *Scandinavian Journal of Work and Environmental Health*, 12:439-447, 1986.

37. Mooney, V. Why do workers' compensation costs keep growing and how can we change this? *Journal of Disability* 3:101-109, 1993.
38. Acute Low back problems in adults: assessment and treatment. *Clinical Practice Guidelines*, No. 14. Rockville, M.D. U. S. Dept. of Health and Human Services, Public Health Service, Agency for Health Care Policy and Research, 1994.
39. Bigos, S.J., Battie, M.C., Spengler, D.M., Fisher, L.D., Fordyce, W.E., Thansson, T.H., Nachemson, A.L., and M.D. Wortley. A prospective study of work perceptions and psychosocial factors affecting the report of back injury. *Spine*, Volume 250, January 1991.
40. Nachemson, A.L. Work for all. *Clinical Orthopaedics and Related Research* 179:77-82, 1983.
41. White, A.A., III, and S.L. Gordon. Synopsis: Workshop on idiopathic low back pain. *Spine* 7:141-9, 1982.
42. Donelson, R.G., Silva, G., and K. Murphy. Centralization phenomenon: Its usefulness in evaluating and treating referred pain. *Spine* 15:3-15, March 1990.
43. Report to the legislature: Health care costs and cost containment in Minnesota workers' compensation. St. Paul, MN. March 1990. pp 1-105.
44. Spitzer, W.O. Scientific approach to the assessment and management of activity-related spinal disorders: A monograph for clinicians. Report of the Quebec Task Force on Spinal Disorders. *Spine* 12:S1-S59, 1987.
45. Cady, L.S., Bishoff, D.P., and E.R. O'Connell. Strength and fitness and subsequent back injury in fire fighters. *Journal of Occupational Medicine* 21:269-272, 1979.
46. Mayer, T.T., Gatchel, R., Mayer, H., Kishino, M.D., and V. Mooney. A prospective two year study of functional restoration in industrial low back injury. *Journal of American Medical Association* 250:450-462, 1987.
47. Waddell, G. A new clinical model for the treatment of low back pain: 1987 Volvo Award in Clinical Sciences. *Spine* 12:632,1987.
48. Brisson, P.M., Nordin, M., and C. Zetterberg. The musculoskeletal system and occupational syndromes. *In:* Rom WN (ed), *Environmental and Occupational Medicine*, 2nd Ed. Boston: Little, Brown and Co., 1992.
49. Klein, B.P., Jensen, R.C., and L.M. Sanderson. Assessment of workers' compensation claims for back strains/sprains. *Journal of Occupational Medicine* 26:433-448, 1984.
50. Venning, P.J., Walter, S.D., and L.W. Stitt. Personal and job-related factors as determinants of incidence of back injuries among nursing personnel. *Journal of Occupational Medicine* 29:820-825, 1987.
51. Deyo, R.A., Diehl, A.K., and M. Rosenthal. Reducing roentgenography use: Can patient expectations be altered? *Archives of Internal Medicine* 147:141-145, 1987.

52. Deyo, R.A., Loeser, J., and S.J. Bigos. Herniated lumbar intervertebral disc. *Ann. International Medicine* 112:598-603, 1990.
53. Pederson, P.A. *Journal of the Royal College of General Practitioners* 31:209-216, 1981.
54. Martinelli, T.A., and S.W. Wiesel. Low back pain: The algorithm approach. *Comprehensive Therapy* 17:3-7, 1991.
55. Cullen, M.R., Chernlack, M.G., and L. Rosenstock. Occupational medicine. Medical Progress. *New England Journal of Medicine* 322:594-601, 1990.
56. Deyo, R.A., Rainville, J., and D.L. Kent. What can the history and physical examination tell us about low back pain? *Journal of American Medical Association* 268:760, 1992.
57. Monaghegh, H.A., and R. Nussbaum. Facet syndrome: A prospective and retrospective study. *Orthopedic Hospital Update*, September/October 1986.
58. See note 56.
59. Waddell, G., McCulloch, J.A., Kummel, E., and R.M. Venner. Nonorganic physical signs in low-back pain. *Spine* 5:117-125, March/April 1980.
60. Deyo, R.A., and A.K. Diehl. Lumbar spine films in primary care: Current use and effects of selected ordering criteria. *Journal of General Internal Medicine* 1:20-25, 1986.
61. Boden, S.D., Davis, D.O., Dina, T.S., Patronas, N.J., and S.W. Weisel. Abnormal magnetic resonance imaging scans of the lumbar spine in asymptomatic subjects. *Journal of Bone and Joint Surgery*, American Volume 72:403-408, 1990.
62. Edelman, R., and S. Warach. Medical progress: Magnetic resonance imaging. *New England Journal of Medicine* 328:708-716, 1993.
63. Wiesel, S.W., Tsourmas, N., Feffer, H.L., Citrin, C.M., and N. Patronas. A study of computer-assisted tomography: The incidence of positive CAT scans in an asymptomatic group of patients. *Spine* 9:549-551, September 1984.
64. Deyo, R.A., Diehl, A.K., and M. Rosenthal. How many days of bed rest for acute low back pain? A randomized clinical trial. *New England Journal of Medicine* 315:1064, 1986.
65. See note 62.
66. Volk, G.J., and M.R. Hendrix. The lumbar disc: Evaluating the causes of pain. *Orthopedics* 14:419-25, 1991.
67. Weber H: Lumbar disc herniation: A controlled study with ten years of observation. *Spine* 8:131-140, 1983.
68. Lehmann, T.R., Russel, D.W., and K.F. Spratt. Efficacy of electroacupuncture and TENS in the rehabilitation of chronic low back pain patients. *Pain* 26:277-290, September 1986.
69. Deyo, R.A. Conservative treatment for low back pain: distinguishing useful from useless therapy. *Journal of American Medicine Association* 250:1057-1062, 1983.

70. Andersson, G.B.J., and J.W. Frymoyer. Treatment of the acutely injured worker. *In:* Pope MH, Andersson GBJ, Frymoyer JW, Chaffin DB, (eds): *Occupational Low Back Pain*. Chicago, IL: Mosby-Year Book, 1991.
71. Carette, S., Marcoux, S., Truchon, R., Grondin, C., Gagnon, J., Allard, Y., and M. Latulippe. Controlled trial of corticosteroid injections into facet joints for chronic low back pain. *New England Journal of Medicine* 325:1002-7, October 1991.
72. See note 70.
73. See note 38.
74. Haldeman, S. Spinal manipulative therapy: A status report. *Clinical Orthopaedic and Related Research* 179:62-70, October 1983.
75. Meade, T.W. Effectiveness of chiropractic and physiotherapy in the treatment of low back pain: A critical discussion of the British Randomized Clinical Trial (letter; comment). *Journal of Manipulative and Physiological Therapeutics* 14:144, 1991.
76. See note 44.
77. Matheson, L.N. Work hardening for patients with back pain. *Journal of Musculoskeletal Medicine* 10:53-63, 1993.
78. Brandon, T.L., and L. Snyder. Work recovery center Piedmont Hospital. *In:* Ogden-Niemeyer L. Jacobs K (eds): *Work Hardening: State of the Art*. Thorofare, NJ: Slack Inc., 1989.
79. Ellexson, M. S.T.E.P.S. Clinic Schwab Rehabilitation Center: *In:* Ogden-Niemeyer L, Jacobs K (eds): *Work Hardening: State of the Art*. Thorofare, NJ: Slack Inc., 1989.
80. Fortenbach, M. The industrial rehabilitation program at the Massachusetts General Hospital *In:* Ogden-Niemeyer L, Jacobs K (eds): *Work Hardening: State of the Art*. Thorofare, NJ: Slack Inc., 1989.
81. Fulper, K.E. Work assessment and rehabilitation program. *In:* Ogden-Niemeyer L, Jacobs K (eds): *Work Hardening: State of the Art*. Thorofare, NJ: Slack Inc., 1989.
82. Reuler, J.B. Low back pain. *Western Journal of Medicine* 143:259-265, 1985.
83. Lavender, S.A., and G.B.J. Andersson. Ergonomic principles applied to the lumbar spine. *Journal of Disability* 3:1-15, 1993.
84. See note 59.

8

Symptom Prevention

"Example is not the main thing in influencing others. It is the only thing."
—*Albert Schweitzer*

Chapter Objectives

After reading this chapter, the reader will:

- Understand the controversies and current recommendations regarding splints, back supports and CTD prevention exercises.

- Be able to describe the basic elements of cumulative trauma disorder prevention at home.

- Know current concepts about computer use.

- Be aware of software sources and the Internet as a resource regarding CTDs.

Case Study

Attitude Adjustment

You observe an employee bending over at the waist while picking up a bed pan that pediatric hospital patients were using as a hockey puck (It was empty, fortunately). You know that he has just returned from a lost-time back injury, his third. You advise him of proper bending and lifting techniques, but he says, "Nurses just have to work with pain." What should you do?

Splints

The use of wrist splints as a CTD prevention and treatment option has become increasingly popular. Research has implicated severe wrist angles as a cause of CTDs.[1] Sometimes splints are used incorrectly. Cock-up splints with the wrist in 30 degrees of extension are often prescribed for carpal tunnel syndrome instead of a neutral splint. A neutral environment can be created by removing the metal bar from such splints and straightening them. Splints are sometimes used in place of proper ergonomics. This strategy may work only in the short run and ignores more fundamental CTD risk factors. Lack of compliance is another problem that arises with the use of splints. Unfortunately, wrist splints are only used 75 percent of the time they are recommended—often due to discomfort experienced by the wearer. Even in a group of 130 patients who were trained about the benefits of splints only 70 percent complied.[2]

About Computers

Who can be affected by cumulative trauma disorders and what are some ways to reduce risk? Computers have been alternatively described as a cause of cumulative trauma disorders and a cure for some of the tasks that lead to CTDs. The use of computers and their concomitant automation have been implicated in cumulative trauma disorders. Excessive keyboard work without breaks can lead to

CTDs. Methods to create user-friendly computer work stations are reviewed in Chapter 5. Computers also offer opportunities to reduce cumulative trauma disorders. How so?

Computerized voice dictation systems offer an almost hands-off approach to word processing. Workers can use speech recognition software to create, edit, print, and send documents by voice. Sections of this book were initially created on such a system. Other software can prompt computer users to exercise and take breaks at regularly scheduled intervals or after a certain number of key strokes and to keep a log of such prevention activities for OSHA purposes.

Voice recognition systems are listed in Table 8.1

Table 8.1 Voice Recognition Systems

System	Contact Information
IBM voice type dictation	1-800-TALK2ME
Dragon Systems	617-965-5200
Kurzweil Applied Intelligence	617-893-5151
Hark System	617-873-4636

Split and adjustable split keyboards are another method to relieve CTDs in the upper extremities. A study comparing split keyboards with standard keyboards showed improved comfort. However, a period of adaptation was necessary since typing speed was reduced and there was an increase in errors.[3]

Supports Under Supported

Back supports are widely prescribed. One grocery store chain requires all workers to wear back supports, although not necessarily around the back. Although workers report a feeling of better support while wearing back supports, their use does not significantly reduce injuries or pain levels. Furthermore, such supports may lead to a false sense of security, are often improperly worn, and their repeated use may reduce abdominal and back strength. They do not address the underlying ergonomic issues, and concern exists that they increase pressure on abdominal organs and may have adverse cardiovascular affects.[4] Finally, no improvements are evident in posture and strength, though Sherman found some increase in spinal compression.[5]

In 1994, the National Institute for Occupational Safety and Health (NIOSH) reviewed the back support issue and found mixed evidence for supports in the literature. NIOSH concluded that while supports do stabilize the spine, they do not improve capacity for lifting.[6] If such supports are used, a job simulated exercise program should be added to improve flexibility, strength, and endurance.[7]

Cost Effectiveness of Back Belts

Surprisingly, one study revealed that the expense of injury among workers using a belt was higher than among those who did not.[8]

CTD Exercises

Does exercise help prevent cumulative trauma disorders? Many health care providers and workers are convinced that exercises reduce the occurrence of cumulative trauma disorders. Conclusive evidence is lacking. The Silverstein study found no significant difference in CTDs between those who did exercises and those who did not.[9] Sometimes exercises are not appropriate to the condition of the worker and incorrect exercises are at times prescribed. Other exercises are embarrassing or disrupt the work. Many workers would be unlikely to comply with an exercise regimen according to the NIOSH review.[10] The Institute reviewed exercises from books and pamphlets designed to prevent cumulative trauma disorders among video display terminal and computer operators and found that 40 percent of exercise programs posed hazards to safety or had the potential to worsen CTDs.[11]

To Exercise or Not to Exercise

In summary, exercises to prevent cumulative trauma disorders are controversial. Most studies indicate limited effectiveness in reducing the incidence of cumulative trauma disorders, but an enhanced sense of well-being. Objectively, CTD exercises may make little difference; subjectively, some workers and health care providers prefer them. Although exercises specific to CTDs are questionable, general aerobic fitness has been found to be helpful in leading to an earlier return to activity in those with back injuries.[12]

CTD Prevention at Home

Principles of cumulative trauma disorder prevention at work are applicable to home activities, but in that environment are unsupervised. The prevention strategies for computers at work apply equally well for a home system, and the techniques to avoid CTDs while mopping, reaching into shelves, vacuuming, and home auto repairs are reviewed in prior chapters (See Table 5.3).

Putting It Together

Many popular cumulative trauma disorder prevention strategies are controversial:

- Preventive exercise efficacy is unproven except for low back injuries.

- Supports and splints can mask underlying causes and lead to deconditioning. If such devices used, wearers require proper training.

- Computers may be the cause, but also the solution for some cumulative trauma disorders. Voice recognition systems such as the one used to prepare part of this book hold promise for decreasing cumulative trauma disorders.

Endnotes

1. Rempel, D., Manoylovic, R., Levinsohn, D., Bloom, T., and L. Gordon. "The effect of wearing a flexible wrist splint on carpal tunnel pressure during repetitive hand activity." *The Journal of Hand Surgery*, Vol. 19 A, 1994.
2. Agnew, M. Compliance in wearing wrist working splints in rheumatoid arthritis. *The Occupational Therapy Journal of Research* 15:165-180, 1995.
3. Ware. Acceptance of the adjustable keyboard. *Ergonomics* 38:1728-1744.
4. Alexander, O. The effectiveness of back belts on occupational injuries and worker perception. *Professional Safety,* Volume 9, 1995.
5. Sherman, B., and J. Woldstad. The effect of a commercially available support belt on torso posture, lift strength and spiral compression. Proceedings of the Human Factors and Ergonomics Society 39th Annual meeting: 605-609, 1995.

6. Lee, K. A review of physical exercises recommended for VDT operators. Cincinnati, OH: National Institute for Occupational Safety and Health, 1995.
7. Genardy, Ash. et al: Can back supports relieve the load of the lumbar spine for employees engaged in industrial operations, *Ergonomics* 38:996-1010, 1995.
8. Mitchell, L.V., Lawler, F.H., Bowen, D., Mote, W., Asundi, P., and J. Purswell. Effectiveness and cost-effectiveness of employer-issued back belts in areas of high risk for back injury. *Journal of Occupational Medicine* 36:90-94, January 1994.
9. Silverstein, B. Can in-plant exercise control musculoskeletal symptoms? *Journal of Occupational Medicine* 30:922-927, 1995.
10. See note 5.
11. See note 5.
12. Cady, L.S., Bishoff, D.P., and E.R. O'Connell. Strength and fitness and subsequent back injury in fire fighters. *Journal of Occupational Medicine* 21:269-272, 1979.

Part III

CTD Programs

"Life is like an umbrella—if we don't learn to work together, we all get wet."

—Heather McNaughton, age 9

Chapter 9: Proposed OSHA Ergonomic Protection Standard 185

Chapter 10: Evaluating the Costs, Cost-Benefit, and Cost-Effectiveness of a Program 193

Chapter 11: Choosing a Consultant 201

Chapter 12: A Sample Program 205

9
Proposed OSHA Ergonomic Protection Standard

"We want to encourage employers to take the high road to safety and we will use our enforcement program to preclude them from taking the low road."

—*Joseph A. Dear, former OSHA Administrator*

Chapter Objectives

Upon completion of this chapter; the reader will:

- Be aware of the basic elements of the proposed OSHA ergonomics standard.

- Know how the OSHA General Duty Clause is relevant to cumulative trauma disorders.

- Be aware of other countries' standards.

Case Study

Another Agenda

Ms. H has been diagnosed with mild tendinitis. In spite of all efforts to change her work station, use of proper splints, and six weeks away from work, she is not at all better. Ms. H asks her doctor "Can't I just go on permanent disability?" She had been previously overheard telling a staff member that she wanted to permanently move to a resort area, to live with her new boyfriend. When the doctor explains that her condition should be temporary, she files a complaint against her company with OSHA.

The Proposed Standard

The proposed OSHA ergonomics standard consists of four key elements:

1. Workplace risk factors correlated with work-related musculoskeletal disorders and the means to decrease those risk factors.
2. Symptoms and signs of musculoskeletal disorders, such as CTD.
3. Worker report of workplace risk factors and work-related musculoskeletal disorders to an identified person, such as a supervisor.
4. Encouragement of early reporting.

The proposed standard includes the following:

1. A workplace risk factor checklist to be completed by the employer.

2. Control measures, with mandates that the employer control problems for each employee exposed to signal risk factors and work related musculoskeletal disorders.
3. An improvement process, which includes a job analysis and control measures for each problem.
4. Manual handling guidelines. Workers should be able to ascertain weight of any item over 25 pounds.
5. Process of task improvement to control CTD.
6. Accurate record keeping, including record of job improvement process.
7. Medical management, including:
   - A musculoskeletal disorder management plan within three days of assessment.
   - Free access to health care providers and an assessment.
   - Assessment of the worker within five work days after signs or symptoms are reported.
   - Periodic walk-throughs by providers.

General Duty Clause

Although OSHA's proposed ergonomics standard failed to pass congressional review, the agency may choose to propose a more narrow standard at some point in the future that will maintain key elements. Meanwhile, it is likely the agency will continue to use its General Duty Clause to fine companies for ergonomics violations.

What is the General Duty Clause (GDC)? The GDC in Section 5 of the OSHAct is the agency's last defense to provide for proper ergonomics in the workplace. The clause is a provision that covers hazards in the workplace not covered by OSHA standards.

Every employer is required to "furnish each of his employees a place of employment which is free of recognized hazards that are likely to cause death or serious physical harm to his employees."[1]

To cite an employer, OSHA must demonstrate that:

- The employer fails to keep the workplace free of a hazardous practice or condition.

- The hazard was recognized and the potential for danger was not addressed.

- There was a substantial probability the condition could lead to serious injury or death.

- There is a practical way to decrease the hazard.

OSHA is not alone. The National Safety Council has a general industry ergonomics guideline that emphasizes a written program, training, employee involvement and management commitment. For a copy, contact National Safety Council customer service at 1-800-621-7619.

The American National Standards Institute (ANSI) Z-365 Committee is actively working on an ergonomics guideline that emphasizes medical management, education and review of work tasks.[2] The federal government is not alone in its efforts to decrease CTDs through proper ergonomics in the workplace, as California is in the process of adopting its own standard. Sweden has developed a voluntary certification program for computers, including keyboards and central processing units.[3] The Canadian province of British Columbia has also discussed CTD regulations.

What to Do if OSHA Inspects?

Inspections are usually prompted by an employee, or ex-employee, complaint or by an accident investigation.

Letters from OSHA will:

1. Describe the allegation.
2. Request information.
3. Inquire about corrective activities, if any.

OSHA inspectors can:

1. Inspect the premises.
2. Take pictures and videotapes. You may request exemption for trade secrets.

Employers may:

1. Request a warrant.
2. Inspect documents, including the complaint.
3. Ask for time to review the above with legal counsel.
4. Escort the inspector, except when employees request a private consultation with the inspector and bring their own consultant at their own expense.

Remember:

1. Keep your OSHA 200 log up to date.
2. Document any CTD training, follow-up, medical management, and surveillance.
3. Act professionally.
4. Contest citations in writing within 15 days, if you disagree with OSHA.

Consider being proactive. Conduct self-audits or request an inspection yourself. OSHA offers a "free look" for employers.

Criticism of the OSHA ergonomics standard has included:

- It was actually a cumulative trauma disorder standard designed to decrease musculoskeletal injuries and not an ergonomics standard.

- Ergonomics involves more than musculoskeletal disorders.

- The standard emphasizes checklists, not job training.

- Decreasing cumulative trauma disorders is an incremental process, slower than the standard would permit.

- Others have gone beyond the requirements suggested by the standard.

- The standard requires a knowledge base and expertise not readily available.

- Work load and work pace are not adequately dealt with in the standard

Unfortunately, OSHA's proposed standard only mandated that workers receive exposure to cumulative trauma disorder prevention practices. Awareness, however, cannot take the place of hands-on practice. Modern learning theory advises learning by doing, not watching.

Putting It Together

- Although legislation on an OSHA ergonomic protection standard was not finalized, it is likely the agency will enforce the OSHA General Duty Clause against violators of ergonomic principles.

- Employers are required to keep their OSHA 200 log up to date and to review it for cumulative trauma disorders.
- Organizations should not, therefore, wait until a standard is adopted by OSHA.
- The OSHA proposed standard was not an ergonomics standard, but rather a cumulative trauma disorders standard.

Endnotes

1. General Duty Clause of the Occupational Safety and Health Act, Occupational Safety and Health Act, Public Law 91-596, 5a, 1970.
2. ANSI Z-365 Release 3rd Draft. *CTD News*, Volume 4, 1996.
3. See note 2.

10

Evaluating the Costs, Cost-Benefit, and Cost-Effectiveness of a Program

"We are not disturbed by things but rather by the view that we take of them. When we meet with troubles let us never blame anyone but rather our opinion about things."

—*Epictetus (60 A.D.)*

Chapter Objectives

After completing this chapter, the reader will understand:

- All programs are not equal—the need to be aware and informed of key elements in choosing and evaluating CTD programs.
- How to determine cost-effectiveness and the elements of a proven program.

- The difference between cost-benefit analysis (deciding whether to do it in the first place) and cost-effectiveness (a comparison of different approaches), and the direct and indirect costs of cumulative trauma disorders.

Case Study

Penny Wise, Pound Foolish

Ms. M is a 24-year-old clerk and an excellent typist (She won the state high school typing contest.) She has had chronic left wrist pain ever since her work area was redesigned. She has seen a hand specialist who told her she has wrist tendinitis and that she should get her work area ergonomics "fixed." She tells her boss, who tells her "It's not in the budget. Just take Percodan if it hurts," which she does! Unfortunately, while under the influence of narcotics, she has a car accident on the way home and is admitted to the intensive care unit.

What are the costs of cumulative trauma disorders? These include direct costs, such as wages and medical care, and indirect costs, such as replacing injured workers, increased productivity costs, management time enrolling and completing workers' compensation paperwork, overtime, training time, administrative costs, and interviewing replacements.

What Should You Do?

- Require someone to receive training in cumulative trauma disorders, or hire a consultant. Training is not a panacea and is ineffective unless combined with other intervention. Safety campaigns require more than putting up a poster. Use local examples that work.

- Recognize that workers must take responsibility for behavioral change. This is difficult at times if job demands inhibit alterations in CTD-prone areas.
- If you hire a consultant, ask how management commitment would be developed.
- Review past efforts, successes, and failures, and determine if your organization has a zero tolerance program for occupational injuries or bonuses for number of days without injuries.
- Ensure that safety audits of workplace hazards have been conducted.
- Ensure that education programs have been implemented to help workers recognize cumulative trauma disorders, and how many lost days, medical expenses, and assess how the number of CTD injuries and illnesses have been related to work.
- Be sure that clear guidelines and information are available for workers when they are hurt, including their benefits, whether to seek care, and availability of modified duty.

Is Management Aware of the Connection Between Gender and Cumulative Trauma Disorders?

In a recent study in Ontario, 55 percent of lost time cumulative trauma disorders claims were among women, who made up 50 percent of the workers in that province. Why? Women often go after work to their second job, which consists of home activities that may also be repetitive, such as ironing, laundry, and folding clothes.[1] There are also certain anatomical differences, such as a smaller wrist diameter. This may contribute to the greater incidence of carpal tunnel syndrome in women.

Costs

What do CTDs cost? Some expenses are obvious, including medical care and lost wages. Others, such as indirect expenses, are not. These include administrative time (filling out forms and making calls), replacement costs for temporary workers, overtime, and training, equipment costs for repair and cleanup, and decreased productivity. In addition, ergonomics and cumulative trauma disorders training may actually increase costs initially. Figure 10.2 should assist you in calculating the total cost of CTDs in your organization.

Cost-benefit versus cost-effectiveness. Cost-benefit and cost-effectiveness are often used interchangeably—and incorrectly. Cost-benefit is an analysis of whether an organization should or should not proceed with a plan, program, or purchase. For example, if company X should implement a CTD prevention program or not is a cost-benefit question.

- If the decision is to proceed with the program, what are the costs (expenses of programs and benefits) versus decreased CTDs?

- What are the costs of not proceeding (perhaps more CTDs, but no expense for consultants, modified workstations, or other preventive measures).

Cost-effectiveness involves a decision based on an analysis of which CTD program would be more beneficial, usually within a given budget.[2] For example, CTD Experts, Inc. will charge one dollar per employee per month for their program with added costs of any ergonomic alterations, such as wrist rests. You have 100 employees. CTD Consulting, Inc., another consulting firm, will charge a flat fee of $1,000.00 and help your facilities department create most of the ergonomic interventions needed. Figure 10.1 will help compare consultants' costs.

	CTD Experts, Inc.	CTD Consulting, Inc.
Consultation charge:		
one time/each time:	_____	_____
per month:	_____	_____
per hour:	_____	_____
Intervention costs		
Equipment	_____	_____
Other costs	_____	_____
Training fees:	_____	_____
Epidemiologic study:	_____	_____
Subtotal:	_____	_____
Discounts:	_____	_____
Finance charges:	_____	_____
Grand Total:	_____	_____

Figure 10.1 Consultant Comparison Chart

The key factor is the payback period for your program—the time necessary to reap the benefits of conducting a CTD program,[3] which can be calculated in Figure 10.2.

Costs	Hours x Pay Rate
Indirect costs:	
• Replacing injured workers, includes:	
Overtime	_____ x _____
Employee turnover	_____ x _____
Decreased productivity	_____ x _____
Use of temporary workers	_____ x _____
Training new hires, temporary workers	_____ x _____
• Paperwork and telephone time	_____ x _____
Completing workers' compensation forms	_____ x _____
Completing disability forms	_____ x _____
• Telephone and other contact time with injured worker, insurance carrier, worker's supervisor	_____ x _____
Total Indirect Costs (add all above) =	_____ x _____

Figure 10.2 CTD Cost Calculation Form

Direct costs

- Legal, if self-insured or a third-party suit _____

- Increased workers' compensation premiums _____

- Injured workers' wages

Medical care if self-insured (limited by stop-loss and catastrophic insurance) _____

Equipment

- Damage _____

- Repair _____

- Replacement _____

Ergonomic alterations and equipment _____

- Property damage _____

- Cleanup expense _____

Total Direct Costs = _____

Grand Total (Indirect and Direct) = _____

Figure 10.2 (cont'd) CTD Cost Calculation Form

Putting It Together

- Use consultants judiciously.
- Most ergonomic changes are common sense; others may require special expertise.
- Usually, workers are the best resource.
- Cumulative trauma disorders are the result of multiple causes. Successful programs, therefore, usually have many components, including training and management commitment.
- Remember that user-friendly work is more than a checklist of physical characteristics. It involves work load and pace, rest rates, and job stress.
- CTD interventions may take months, even years. Do not abandon a worthwhile program if results are not evident in a few weeks or months time.

Endnotes

1. Ashbury, F. Occupational repetitive strain injury and illness in Ontario, 1986 to 1991. *Journal of Occupational and Environmental Medicine* 37:79-85,1995.
2. Eddy, D.M. Clinical decision making from theory to practice: cost effectiveness analysis: Conversation with my father. *Journal of American Medical Association* 267:1669-75,1992.
3. Oxenburgh, M. *Increasing Productivity and Profit through Health and Safety*. Chicago, IL: CCH International, 1991.

11

Choosing a Consultant

"May we have the foresight to see where we are going and the hindsight to see where we have been and the insight to know when we have gone far enough."

Chapter Objectives

After completing this chapter, the reader will know:

- The important elements in choosing CTD consultants and evaluating their success.
- All consultants are not equal. A consumers' guide is offered.

Case Study

Poor Ergonomics Advice

Ms. W is a 38-year-old data entry clerk with a numb finger. An "ergonomist" visited her work site and told her not to use a wrist bar to support her forearms on her keyboard. Within two days, not only was her right little finger numb, but so was her left!

A Consumers' Guide to CTD Consultants

1. Ask prior clients:
   - How well did the consultant meet your expectations? What improvements in productivity and CTD rates occurred?
   - How well did the consultants work with your employees—as subjects or as partners? To what degree were employees allowed to participate?
   - If you were to change one thing about the consultants work, what would it be?
   - How well did the consultant perform when things went wrong? What did they do about it?
2. Ask other vendors, unions, employers, and professional or trade organizations similar to yours about the consultant's track record.
3. Ask the consultant:
   - What is their level of expertise in the field (This varies greatly)? Have they worked in your industry? How do they measure their success?
   - Are they an ergonomics trained physical therapist, occupational therapist, occupational health nurse or industrial hygienist or a Certified Professional Ergonomic Evaluator (CPEE), Certified Professional Ergonomist (CPE) or board certified by the American Board of Medical Specialities (e.g., Occupational Medicine)? Be cautious of "ergonomic" or ergonomically friendly slogans without verifying past experience. Some consultants may promote their ergonomics expertise when they are actually specialists in other arenas, such as confined space issues.[1]

- Can you see past reports, recommendations, and a sample program?
- Have they ever lost a client?
- What did they do when things did not go well?
- How familiar are they with your work process?
- What role will your staff, including low-level workers, play in the consultation?
- What training programs are offered?
- Do they have easy-to-follow checklists?
- What evaluation strategies are available to measure success or outcomes?
- Do they carry errors and omissions insurance in case you are damaged by their advice (although this is probably not necessary for small projects), and workers' compensation insurance (in case they slip while doing a walk-through)?
- What CTD hazards do they observe during their walk-through?
- Are they available for other consulting such as hearing protection?
- What follow-up do they offer (e.g., three month revisit)?
- How much money will their program save?

Putting It Together

- Consultants, if carefully chosen, can help move your CTD prevention and control project forward.

- It pays to ask around. Past clients are the best—and worst—references.

- Certification by the Board of Certification in Professional Ergonomics is not a requirement, but all things being equal, it indicates an important degree of training.

- Checklists are helpful, but they should never replace knowledge of the process, jobs, and work tasks.

Endnote

1. Koutsandreas, Z.T. "How to choose an Ergonomic Consultant." *Workplace Ergonomics*, Volume 2, 1996

12

A Sample Program

"Never doubt that a small group of thoughtful, committed citizens can change the world; indeed, it is the only thing that ever has!"
—Margaret Meade

Chapter Objectives

After completing this chapter, the reader will understand:

- How to develop a sample CTD program.
- The nominal group technique as a program initiation strategy.

Case Study

An integrated health care organization experienced a 50 percent increase in cumulative trauma disorders (CTD) over a two-year period. Cumulative trauma disorders were one of the leading causes of morbidity and absenteeism in the workplace, and preventive programs were seen as a method of reducing the incidence and severity of these conditions. In a training *program aimed at reducing CTDs in the work-place, 184 managers in a private health care organization, consisting of three hospitals, ambulatory clinics, and a nursing home, participated in a series of seminars. These seminars focused on three areas of CTD concern: 1) causes; 2) early recognition; and 3) prevention. Using sample workplace settings, each manager then demonstrated the knowledge and skills needed to:*

- *Evaluate work sites.*
- *Assess ergonomic hazards.*
- *Correct hazards.*
- *Understand the causes of CTD such as poor ergonomics.*
- *Learn how to prevent CTDs through early recognition and other preventive measures.*

An Approach to CTDs as a Management Problem

Corporate and individual management of CTDs in an occupational setting presents complex social and medical issues. Most cases of CTD, including back and shoulder sprains and strains, tendinitis, bursitis and carpal tunnel syndrome (CTS) can be related to a lack of sound ergonomic techniques applied to the workplace. We, the authors, have found that the prevention and control of CTDs in the workplace benefits from a collaborative approach, with team members representing management, employees, and health care personnel.

A model of primary, secondary, and tertiary prevention seems appropriate to structure management response to the problem of occupational CTDs. Primary prevention measures can target CTD risk factors in the workplace in an effort to reduce the incidence of CTD symptoms, before disease occurs. Such an approach will be outlined in the following case.

Methods

CTDs are of obvious interest not only to health care professionals who treat injured workers, but also to managers who wish to prevent these injuries. To determine if educating managers in sound ergonomic principles in the workplace could have an effect on the number of CTD cases, the authors developed and instituted the training program described below for a large employer.

Training was given to 184 managers as part of a required CTD prevention program for a private health care organization with a broad spectrum of workers, including maintenance, loading dock, food service, housekeeping, laboratory, nursing, and rehabilitation services—all areas that were experiencing CTDs. Managers met in groups of 10 or 12. Eight seminars of four hours in length were held for each group over a one-year period.

A multidisciplinary approach provided instruction by a safety specialist, physical therapist, occupational medicine physicians, and human resources personnel. Managers were instructed how to analyze the work site and work stations to identify potentially hazardous work postures. They learned how to modify poorly designed workstations, such as by adjusting the work to an employee's range of motion, using postural reliefs, lifts, document holders, footrests, arranging keyboards to allow wrists to remain in a neutral position, orienting computer display terminals to avoid cervical strain and adjusting chairs so the hips are at 90 degrees of flexion. Other preventive measures were taught, such as rotating tasks to avoid constant repetition and teaching workers to engage in simple stretching exercises, such as shoulder rolls and neck rotations for typists.

Practical training, using a case study and hands-on exercise approach, was chosen as an effective tool in preventing CTD in the workplace. Exercises on the evaluation of CTD were employed to redesign sample worksites and implement ergonomic changes. After completing the training and the cases, each manager had a basic competency in preventive principles in sample work settings. In addition, each manager was required to demonstrate individual competency in ergonomic principles to the course instructors.

Results

We learned to assume nothing. We found individuals who did not know how to use the adjustment knob on chairs. A course work evaluation form completed by each manager showed they believed all three objectives were successfully met. 184 managers completed the evaluation questionnaire. On a 1-to-5 Likert-type scale (1=unsatisfactory, 2=poor, 3=adequate, 4=good, 5=outstanding), exercises were rated at 4.67, course organization was rated at 4.67, and overall course assessment was 4.68. Questions used to elicit the numerical response concerned how effective the presentations were from a learning standpoint, the value and applicability of the course for their particular workplace, and the degree of active participation in the learning process.

During a one-year period following training, the incidence of CTD in the work settings of the trained managers decreased 9 percent. Cost of treatment declined 19 percent, and lost days due to CTD fell 26 percent. Confounding factors, such as turnover in the study population and changes in job duties, remained relatively constant. Information was not collected on non-work-related CTDs.

This program presents a method of decreasing CTDs in the workplace through a management training program in response to a problem of increasing occupational CTDs. The strategy was to target risk factors in the workplace for a large employer and reduce those risks to decrease the number of cases of CTD.

Table 12.1 illustrates some of the team problem-solving areas. Similar cases were used during the training course.

Table 12.1 Team Problem-Solving Areas

Area 1	
Job/tasks:	Salad preparer—hospital cafeteria.
Risks/hazards:	Worker used repetitive hand and wrist movements to cut and hold vegetables and a constant grip to hold the knife.
Changes/improvements:	A vegetable dicer was mounted which eliminated the use of a knife.
Cost:	$100.00
Area 2	
Job/tasks:	File clerk.
Risks/hazards:	Worker reached above head to file heavy folders, extending neck and elevating arm.
Changes/improvements:	A step stool was provided.
Cost:	$24.00

Conclusion

A model aimed at primary prevention of CTDs seems appropriate to form a management and employee approach to respond to the problem of occupational CTDs. Targeting risk factors in the workplace and reducing those risks can decrease the number of cases of CTDs. Practical training, using a case study and hands-on exercise approach, can be an effective tool in preventing CTDs in the workplace.

Key elements of an effective program include:

- Employee involvement (self-responsibility model).
- A preventive focus. It is less expensive and more effective than treatment in the effort to alleviate the problem of occupational CTDs.
- Management commitment.
- Written program.
- Training.
- Checklists (These should not replace knowledge of the process, jobs, and tasks).

Remember to:

1. Create CTD education programs for all employees, with mandatory attendance by managers so that all are able to recognize CTD hazards and understand prevention of their occurrence.
2. Recognize that a CTD prevention program has no end. The process, like education, is continual.
3. Develop a CTD management program, with an achievable goal. Zero occupational injuries and illnesses may be an unrealistic goal, but a ten percent decrease each year may be realizable.

4. Create an open door policy so that employees can express their concerns about CTDs to management.
5. Perform regular CTD audits of workplace hazards—CTD "hot spots" or areas of known hazards.
6. Analyze your injury and loss information to learn from the present to correct the future.

If a CTD occurs:

1. Ensure that an injured employee receives prompt, quality care.
2. Report CTD cases to your third-party administrator and/or workers' compensation insurer.
3. Direct workers with CTDs to providers experienced in their management.
4. Focus on function. Consider what the worker can safely perform, and what modifications in duty and hours are advisable.
5. Explain to employees what their benefits will be, and tell them that their new job assignment is to get better by keeping their appointments and adhering to their rehabilitation plan.
6. Keep in frequent contact with employees. Contact should be at least weekly to decrease their feelings of isolation or alienation.
7. Communicate with providers about a return-to-work date on modified duty. If the worker is temporarily unable to perform usual job duties, search for suitable alternative work.
8. When injured workers are performing in another department, make certain that they understand they are still responsible for attendance and quality of work.

9. Training to prevent CTDs is not a one-time event and is ineffective unless combined with other interventions. As a comparison, a seat belt campaign requires more than putting up a poster.
10. Search for local examples that work.
11. Recognize that workers must take self-responsibility for behavioral change. This is difficult if job demands inhibit alteration in CTD-prone behaviors.
12. Use modern management tools to develop your own programs and prioritize areas of CTD concern and possible action.
13. Record CTD cases on your OSHA 200 log.

The Nominal Group Technique

The nominal group technique (NGT) is one tool that can generate ideas quickly by brainstorming, and can survey the opinions of a small group of 10 to 15 people.[1] Why use NGT? It generates ideas and suggests actions in a short period of time. Its advantages include:

- It focuses on problems, not personalities.
- It keeps communication open.
- It fosters participation by "wall flowers" who might have good ideas that never get heard.
- It allows conflicting ideas to be voiced.
- It builds commitment and consensus.

Figure 12.1 summarizes the NGT method.

> 1. Present the issue and the following instructions. An example of an issue is: Why do we have so many CTDs?
>
> 2. Allow participant to generate ideas with no discussion.
>
> 3. Survey the ideas round robin style from each participant, with no more than one idea at a time. Write ideas on blackboard or flip chart. Continue until all ideas are listed.
>
> 4. Process and clarify ideas. Eliminate duplicates. Combine similar ideas. Limit discussion to brief explanations or analysis.
>
> 5. Focus on clarification, not debate.
>
> 6. Allow participant to establish individual priorities silently.
>
> 7. Add up votes.
>
> 8. Work on an action plan using the same strategy. You might take the top five from six, then develop a plan for those items. This is often necessary in an area of limited resources.

Figure 12.1 Conducting a NGT Session

Putting It Together

- When developing a program, use consultants judiciously. Many ergonomic changes are common sense, while others do require special expertise.
- Usually workers are the best resource.
- CTDs are the result of multiple causes. Successful programs have many components including training, practice, and work site changes.
- Remember that user-friendly work is more than a checklist of physical findings. It involves workload, work pace, rest rates, and job stress.

Endnote

1. Van de Von and Gustafson. *Group techniques for program planning.* Glenview, IL: Scott Foresman and Company, 1975.

Glossary

abduct — To move or draw away from the axis of the body or limb.

ACGIH — American Conference of Governmental Industrial Hygienists; the group that establishes threshold limit values and biological exposure indices for chemical substances and physical agents in the workplace.

acromion — Lateral extension of the spine of the scapula, projecting over the shoulder joint and forming the highest point of the shoulder.

acromioclavicular — Pertaining to the acromion and clavicle, especially to the articulation between the acromion and clavicle (see separate definitions).

ADA — Americans with Disabilities Act, a federal law passed in 1990, which prohibits discrimination against persons with disabilities.

adduct — To move or draw closer to the axis of the body.

adhesive capsulitis	Inflammation of the joint capsule of the shoulder and the subdeltoid bursa, characterized by shoulder pain of gradual onset of pain with increasing pain, stiffness, and limitation of motion; also called adhesive bursitis and frozen shoulder.
ankylosing spondylitis	Form of rheumatoid arthritis that affects the spine; causes inflammation of the spinal joints, pain and stiffness.
ANSI	The American National Standards Institute is a standard making organization.
anterior interosseous	Complex of symptoms caused by a lesion of the anterior interosseous nerve, usually resulting from fracture or laceration, but sometimes resulting from compression, with pain in the proximal forearm and weakness of the muscles innervated by the nerve.
anthropometry	Science that deals with the measurement of size, weight, and proportions of the human body.
antiglare screen	Screen with a special coating to prevent the reflection of light.
bicipital tendinitis	Inflammation of the biceps tendon.
biomechanics	Application of mechanical laws to living structures.

BLS	Bureau of Labor Statistics; the division of the Department of Labor that records statistics related to occupations and the workplace.
bursitis	Inflammation of a bursa, a saclike cavity, especially located between joints or at points of friction between moving structures.
cauda equina	Collection of spinal roots that descend from the lower part of the spinal cord and occupy the vertebral canal below the cord; their appearance resembles the tail of a horse.
clavicle	Forms the anterior portion of the shoulder girdle on either side; also called the collar bone.
computerized axial tomography	Process of producing a CAT scan, computerized cross-sectional images which provide a noninvasive means of visualizing all body structures.
cost-benefit	Relative value of an action based on comparing its worth to its cost.
cost-effectiveness	Producing optimum results for the expenditure.
CTD	Cumulative trauma disorder; a recurrent or persistent musculoskeletal pain involving soft tissues, usually due to minute, progressive stresses imposed over time.

CTS Carpal tunnel syndrome; a complex of symptoms resulting from compression of the median nerve in the carpal tunnel (the osseofibrous passage for the median nerve and the flexor tendons), with pain and burning or tingling sensations in the fingers and hand, sometimes extending to the elbow.

cubital tunnel syndrome Complex of symptoms resulting from injury or compression of the ulnar nerve at the elbow, with pain and numbness along the ulnar aspect of the hand and forearm, and weakness of the hand.

de Quervain's syndrome Painful tenosynovitis due to inflammation of the common tendon sheath of the abductor pollicis longus and the extensor pollicis brevis, the muscles that move the thumb.

downsizing Making a company or industry smaller by eliminating or simplifying functions or processes and consequently eliminating some jobs.

electromyography Electrodiagnostic technique for recording the extracellular activity of skeletal muscles at rest, during voluntary contractions, and during electrical stimulation; the purpose is to determine the integrity and function of muscle fibers.

endoscope	Instrument for examining the interior of a body canal or hollow organ.
epicondyle	An eminence upon a bone to which is attached a tendon.
epicondylitis	Inflammation of the epicondyle or of the tissues adjoining the epicondyle of the humerus.
epidural injection	An injection of a medication, such as a steroid, at the outside of the dura mater lining of the spinal cord.
ergonomics	Science relating to man and his work, and concerned with optimally and safely fitting the human to the work by using anatomic, physiologic, and mechanical principles.
essential function	Fundamental part of job duties; to eliminate it would change the job significantly.
facet syndrome	Inflammation of vertebral facet joints, joints composed of planar articular surfaces resulting in low back pain.
FCE	Functional capacity evaluation; an assessment of what a worker physically can or cannot do; the evaluation defines what someone is physically capable of performing.

flexion	Act of bending, or the condition of being bent.
frozen shoulder	Another name for adhesive capsulitis; see that definition.
functional capacity	What a worker can or cannot do from a physical standpoint.
functional job analysis	An evaluation of the functions of a job and its demands.
General Duty Clause	Part of the Occupational Safety and Health Act of 1970 that ensures a safe workplace for every worker.
hand-arm vibration syndrome	Caused by injury to the fingers and hands of workers who use vibrating hand tools; symptoms include tingling, numbness, and blanching of the fingers with probable loss of muscle control and reduction of sensitivity to heat and cold with accompanying pain on return of the circulation; sometimes referred to as HAVS.
herniated nucleus pulposus	Protrusion of the semifluid mass of Fine white and elastic fibers that forms the central portion of an intervertebral disk.
job analysis	An assessment of components of a particular work task and its inherent risks.

job rotation	Allowing workers to alternate job duties with other workers, so that one person is not performing the same job functions all the time.
lumbar sacral syndrome	Low back pain due to disease or trauma at these sites; the lumbar region refers to the part of the back and sides between the lowest ribs and the pelvis; sacral refers to the sacrum, a triangular bone made up of five fused vertebrae and forming the posterior section of the pelvis; lumbar sacral refers to both the lumbar region and the sacrum.
lumbar supports	Devices for lending stability to the lower back, such as a chair or cushion designed for that purpose; other supports are back belts or corsets.
magnetic resonance imaging	Noninvasive diagnostic procedure to obtain detailed sectional images of the internal structure of the body.
marginal function	Job duty that is peripheral to the main purpose of the job.
median nerve	Nerve that supplies sensory and motor function to the flexor muscles in the arm and hand and sensory innervation to the corresponding skin.

neoplasm — Any new abnormal growth, either benign or malignant; specifically, a new growth of tissue in which the growth is uncontrolled and progressive; also called a tumor.

nerve entrapment — Compression of a nerve due to various causes, such as fracture, tumor, scar tissue, or fluid buildup.

nominal group technique — Method of group discussion used to generate ideas and suggest solutions in a short amount of time; also referred to as NGT.

nonsteroidal anti-inflamatory drug — Drug, which is not asprin or steroids, used to reduce inflammation and therefore pain.

OSHA 200 log — Accumulated record of work-related injuries and illnesses kept by the employer; "200" is the form number of the log sheets. Compare OSHA 101, which is the form used to report separate individual cases and the details of these work-related injuries or illnesses.

paresthesia — Abnormal or impaired skin sensation, such as burning, prickling, itching, or tingling.

pinch grip — Grasping objects with two fingers.

postoffer examination	Medical examination carried out on a worker after an offer of employment has been made.
power grip	Grasping objects with the entire hand.
preplacement examination	Same as postoffer examination; (*see* postoffer examination).
pronate	To turn (the palm or inner surface of the hand or forelimb) downward.
pronation	Act of turning the palm or inner surface of the hand or forelimb downward.
pronator syndrome	Disorder caused by entrapment of the median nerve by the pronator teres muscle in the forearm; this results in pain and weakness in the forearm.
psychosocial factor	Pertaining to or involving both psychological and social aspects.
radial tunnel	Narrow passage through which the radial nerve traverses; radial pertains to the radius of the forearm or to the radial (lateral) aspect of the arm as opposed to the ulnar (medial) aspect.
Raynaud's phenomen	Intermittent bilateral attacks of ischemia of the fingers or toes, marked by severe pallor, and often accompanied by paresthesia or pain;

it is brought on characteristically by cold or emotional stimuli and relieved by heat, and is due to an underlying disease or anatomical abnormality.

reasonable accommodation Measures, required to ensure equal opportunity for disabled workers, which an employer can reasonably allow such that they do not adversely affect the business (not costly, extensive, substantial, or fundamentally alter thhe nature of the workplace). These include: making existing facilities accessible to individuals with disabilities; job restructuring; modified work schedules; reassigning workers to vacant positions; modification of equipment or acquiring of new equipment; adjusting of exams, training materials, or policies; and providing qualified readers or interpreters.

repetitive motion injury Injury resulting when the same motion is performed over and over many times each day, and the motions eventually exceed the ability to recover from this stress, especially if forceful contractions of muscles are involved in the repetitive motions. (This is the same as cumulative trauma disorder.)

repetitive trauma injury Condition caused by a repeated motion, pressure, or vibration; also known as repetitive strain injury. (This is another name for cumulative trauma disorder).

risk factor An aspect of a worker's history or of the job that makes that individual susceptible to injury or illness.

rotator cuff tendinitis Inflammation of the tendons that support the shoulder joint and merge with the joint capsule; the several tendons combine and become, collectively, the rotator cuff.

sciatica Pain due to involvement of the sciatic nerve; this is the large nerve that supports functions in the lower extremity; the nerve exits from the lumbosacral spine and it, along with its branches, travels the length of the hip, leg, and foot. Pain can be anywhere along this route.

signal risk factor Features of the workplace associated with an increased possibility of cumulative trauma disorders, as determined by OSHA; these are listed in Figure 3.2.

sit/stand chair chair that allows a worker who sometimes sits and sometimes stands at his work to sit at a greater height or to stand in front of the chair; this eliminates reaching to the work surface or sitting to a greater depth, as with a conventional chair.

supraspinatus tendinitis   Inflammation of the supraspinatus tendon; this is one of the tendons making up the rotator cuff; another name for rotator cuff (*see* rotator cuff).

syndrome   Group of symptoms or signs that collectively indicate or characterize a disease or other abnormal condition.

tendinitis   Inflammation of tendons or tendon-muscle attachments; a tendon is a fibrous cord by which a muscle is attached.

tenosynovitis   inflammation of a tendon sheath; the tendon sheath is a fibrous covering over a tendon.

third party administrator   Insurer engaged by the employer to coordinate and pay for claims.

thoracic outlet syndrome   Numbness, tingling, and pain in the shoulder and arm, especially when the shoulders are pulled back and the affected arm is raised; results when the nerves of the brachial plexus and brachial artery branches are pressed between the muscles of the shoulder and the neck; called TOS.

ulnar nerve entrapment Entrapment of the ulnar nerve that runs along the ulna; leads to ulnar nerve syndrom (see nerve entrapment); the ulna is the bone extend-ing from the elbow to the wrist on the side opposite to the thumb.

ulnar nerve syndrome Pain and weakness that results from ulnar nerve entrapment (*see* nerve entrapment).

VDT Visual or video display terminals; both terms refer to the screen at a computer station.

vibration-induced white finger Tingling and numbness in the fingers after using a vibrating tool; attacks can be aggravated or induced by cold temperatures.

vibrometer Device that measures the amplitude of vibrations.

vibrometry Measurement of vibrations.

walk-through An inspection of a work site by safety and health professionals for the purpose of detecting and correcting potential hazards.

whole-body vibration vibration transmitted to the entire body through some supporting structure, such as a vehicle seat or a building floor; vibration can be defined as a back-and-forth, up-and-down, side-to-side linear motion that emanates from and returns to some defined position.

work hardening Process by which the worker is prepared for return to work after an injury through gradual use of affected muscles, joints, and nerves in a simulated work setting; restoring the function of these muscles, joints, and nerves is called reconditioning.

workstation Work or office area assigned to one person, often one accommodating a computer terminal or other electronic equipment.

workers' compensation Insurance required by law from employers for the protection of employees while engaged in the employer's business; this insurance pays medical bills, disability, and lost income to the employees in the event of a work-related injury.

Index

A page number followed by an "f" or a "t" denotes a reference to a figure or table.

A

Accommodations
 ADA requirements for, 65-70
 cost of, 66
 designing, 72f
 Job Accommodations Network, 66
 low cost alternatives, 110, 110t
 reasonable, 76
 rules for, under ADA, 74f
Acetaminophen, for lower back pain, 153f
ACGIH. *See* American Conference of Governmental Industrial Hygienists
Acromegaly, and risk of CTDs, 147f
Acromioclavicular syndrome, 35, 36t, 139t
 diagnosis and treatment, 149
 risk factors for, 147f
Activity, return to, 156
Acupuncture, for lower back pain, 153f
ADA. *See* Americans with Disabilities Act
Adhesive capsulitis, 139t
 diagnosis and treatment, 149
 risk factors for, 147f
Adson's test, 36f, 148
Agency for Health Care Policy and Research (AHCPR), 153
Aging, and CTD risk, 19
AHCPR. *See* Agency for Health Care Policy and Research
Aircraft engine manufacturing, 22
Alcohol abuse, and risk of CTDs, 147f, 161
Allen's test, 151
American Conference of Governmental Industrial Hygienists (ACGIH), vibration guidelines, 48

Americans with Disabilities Act (ADA), 44-45, 62, 65-66, 166, 168
 compliance guidelines, 72f-73f
 effect of, 70-71
 on job application process, 71-76, 72f-74f
 and functional capacity evaluations, 64
Amyloidosis, and risk of CTDs, 147f
Ankle jerk test, 160
Anterior interosseous syndrome, 139t
 diagnosis and treatment, 151
Antidepressants, for chronic lower back pain, 165
Applicants, interview restrictions under ADA, 71-76
Aprons, 53
Arm, workstation design and, 95, 98f-100f
Arm guards, 53
Arm injuries, common forms, 139t
Arm rests, 53
Arthritis, 22
As in Red Hot keyboard, 108, 108t
Assembly line workers, 22, 34-35, 48, 141-143, 148-149
Asymmetric multiplier (AM), in NIOSH lifting guides, 83
Attitude, and risk of CTDs, 65f
Auto industry workers, 2, 12
Auto repair technicians, 148, 152

B

Back disorders, 36t, 139t, 153-168
 bowel/bladder changes in, 163
 chronic, 165
 cumulative nature of, 111

foot drop in, 158, 164
inflammatory vs. noninflammatory causes, 161
laboratory studies of, 161
with leg pain, 164
medication for, 164
neurologic changes in, 161
physical examination for, 158
prevention of, 166
pulse rate in, 160
reflexes in, 160
and renal compromise, 164
risk factors for, 47
testing, guidelines for, 162f
and vibration, 49
work hardening for, 165, 166, 169
Back supports, 126, 133f, 163-164, 166
Back, workstation seating and, 95, 96f-97f
Bed rest, for lower back pain, 153f
Behavior problems, versus medical conditions, 73f
Bending, reduction of, 113t, 130, 132f
Bicipital tendinitis, 34-35, 36t, 139t
diagnosis and treatment, 149
risk factors for, 147f
Bifocals, and monitor position, 41, 100f, 103
Biofeedback, for lower back pain, 153f
Biomechanics
basic principles of, 40
in work site assessment, 122-123
Bladder dysfunction, in lower back pain, 163
Body, positioning of, 47
Bone scans, for low back pain, 161f
Bone spurs, 31
Bowel dysfunction, in lower back pain, 164
Brick layers, 143
Buffers, 142
Bureau of Labor Statistics (BLS), 2, 12
definition of CTDs, 10
Bursitis, 36t, 139t
diagnosis and treatment, 141-142
Butchers, 33, 152

C

Calcium channel blockers, 49
Carpal tunnel syndrome (CTS), 32, 48, 168
causes of, 32
cost of, 20-21
diagnosis and treatment, 143-145, 144f, 146t
Carpenters, 33, 152
Carpet layers, 47
Carrying, reduction of, 104, 131
Cashiers, 41, 140, 143, 148, 152
Cervical strain, 139t
diagnosis and treatment, 140-141
Cervical syndrome, 36t
Chairs
adjustment of, 41
properly designed, 107, 133f
Clerical workers, 16, 22, 40-41, 140-141, 143
Clinical Practice Guidelines Number 14: Acute Low Back Problems in Adults, 153, 153f
Clothing manufacturing, 22
Cold, and risk of CTDs, 49
Committee on Accreditation of Rehabilitation Facilities, 165
Communication, importance of, 156-157
Computerized axial tomography, for low back pain, 161, 162f
Computers, 13, 16, 41, 47. *See* also Clerical workers; Data entry workers; Keyboard operators
ergonomic design of, 46
glasses and, 41, 53, 100f, 103
Conflict with supervisors, and CTDs, 109
Construction workers, 34-35, 141-142, 148-149
Contact stress, and risk of CTDs, 52, 124-125
Control, lack of, and risk of CTDs, 51
Cost, of cumulative trauma disorders, 3, 20-21
Coupling multiplier (CM), in NIOSH lifting guides, 83

Cross straight leg raise test, 160
Crouching, avoiding, 125-126
CTDs. *See* Cumulative trauma disorders
CTS. *See* Carpal tunnel syndrome
Cubital tunnel syndrome, 139t, 146, 147f, 151
  diagnosis and treatment, 151
Cumulative trauma disorders (CTDs)
  administrative solutions for, 54
  biomechanical basis for, 40
  causes of, 11, 13f, 17. *See also* individual professions
    historical, 21-22
  common forms, 36t, 139t
  cost of, 3, 20-21
  cumulative nature of, 111
  definition of, 2, 10
  development of, 28-29
  diagnosis of, 139-167
  incidence of, 2, 11-12, 13f, 16
  prevention of, 65f, 111-113, 113t, 122t
    carpal tunnel syndrome, 144f, 145, 146f
    evaluation checklist, 132f
    exercises for, 42f
    lower back injury, 164-166
    lower extremities, 125-126
    upper extremities, 122-125
    via functional capacity analysis, 114
  reporting of, increased, 19, 20f
  risk factors for. *See* Risk of CTDs
  symptoms of
    common, 31f, 157, 158t
    determining, 157
  treatment of, 139-167
    and diagnostic uncertainty, 154-155
    rapidity of, 3

D

Data entry workers, 140, 148
Degenerative joint disease, 31
Dental workers, 140
de Quervain's disease, 33, 36t, 139t
  diagnosis and treatment, 152
Diabetes mellitus, and risk of CTDs, 147f
Diagnostic uncertainty, 154

Direct threat, definition of, 73f
Disability
  definition of, 72f, 74f
  documentation of, 76
  temporary, 72f
Disk
  herniated, 112, 159
  protruded, 112
Distance multiplier (DM), in NIOSH lifting guides, 83
Drug use, and CTD risk, 20f, 154, 161
Dvorak keyboard, 108, 108t

E

EEOC. *See* Equal Employment Opportunity Commission
Elbows, proper position of, 133f
Electricians, 17
Electric power manufacturing, 22
Electric stimulation, for lower back pain, 153f
Electroacupuncture, for lower back pain, 164
Electrodiagnostic studies, preplacement, 64-65
Electromyography
  for diagnosis of low back pain, 161f
  for numbness, 31
Entitlement attitude, and CTD rate, 19
Epicondylitis, 34, 36f, 139t
  diagnosis and treatment, 141
Equal Employment Opportunity Commission (EEOC)
  accommodations enforcement, 66
  rule and technical assistance booklet, 65
Equipment. *See also* Tools
  maintenance of, 105, 124
  size of, and women, 15
Ergonomic analysis
  of work, 104-105
  of work area, 81-82, 90f, 103-109, 118-122, 132f
  checklist, 132f
  of workers, 105
Ergonomics
  computer design, 46

OSHA standards, 54, 132f-133f, 134
   tool design, 46
   work station design, 85-87, 88f-90f, 92f
Erythrocyte sedimentation rate testing, for lower back pain, 161
Essential functions, 72f
   definition of, 44-45
   samples of, 75f
Ethnic groups, and CTD risk, 22
Examination
   for lower back pain, 159
   post-offer, 62-66, 67f-69f, 70, 73f-74f, 76
   return-to-work, 65
Exercise
   for back pain, 153f, 165
   for prevention of CTDs, 42f
Exertions, forceful, and CTDs, 47
Extremities
   lower, prevention of CTDs in, 125-126
      risk factors for, 47
   upper
      CTDs in, 36t
      musculoskeletal disorders (UEMSD), 48
      nerve entrapment syndromes, 147-151
      risk factors for, 147f
      pain in, 140
Extrinsic factors in CTD injury, analysis of, 61

F

Facet joint injections, for lower back pain, 165
Facet syndrome, 159f, 163
Fatigue, and risk of CTDs, 50
FCE. *See* Functional capacity evaluation
Femoral stretch test, 160
Fever, 161
Finkelstein test, 33, 36f, 152
Fitness, general, and CTD risk, 18-19
Flick test, 36f
Flooring, 104
Foot drop, in lower back pain, 163
Foot rest, 110t

Foot support, 91, 93f-94f
Forceful exertions, and CTDs, 47, 123-124
Forceful manual lifting, OSHA definition of, 44
Ford, Henry, 21
Four-part paper users, 33
Frequency multiplier (FM), in NIOSH lifting guides, 84
Frequent manual handling, OSHA definition of, 44
Froment's paper sign, 151
Frozen shoulder syndrome, 139t
   diagnosis and treatment, 149
   risk factors for, 147f
Functional capacity evaluation (FCE)
   and ADA requirements, 64
   definition of, 113-114

G

Gilliat's test, 145
Glare, and monitor position, 103
Glasses, and computer use, 41, 53, 100f, 103
Gloves, 133f
   types and sizes, 124, 126-127
   vibration dampening, 53
Golfer's elbow (medial epicondylitis), 34, 36f
   diagnosis and treatment, 141
Gout, and risk of CTDs, 147f
Grinder operators, 35, 47, 142-143, 149
Grocery checking, 41, 148

H

Hand-arm vibration syndrome (HAVS), 35, 48, 152
Handling, frequent manual, OSHA definition of, 44
Headaches, 148
Headsets, to reduce neck bending, 53
Head, workstation design and, 103
Heat, for lower back pain, 153f
Heavy lifting, 113t

Heredity, and risk of CTDs, 147f
Herniated disk, 112, 159
Herniated nucleus pulposus (HNP), 158-161
Hips, workstation seating and, 95
History of patient
 determining, 157
 prior, 46
HNP. *See* Herniated nucleus pulposus
Hobbies contributing to CTDs, 53
Horizontal multiplier (HM), in NIOSH lifting guides, 83
Hosiery manufacturing, 22
Hypothenar hammer syndrome, 48
Hypothyroidism, and risk of CTDs, 147f

I

Impairment, fear of, 156
Impingement test, for rotator cuff tendinitis 142
Infection, and risk of CTDs, 147f
Interview restrictions, under ADA, 71-76
Intrinsic factors, in CTD injury, definition of, 61
Inversion therapy, 157

J

Jack hammer operators, 43t, 47-48
Job Accommodations Network, 66
Job analysis, 80-81
Job description, written
 importance of, under ADA, 70
 sample, 75f
Job rotation, to reduce CTD risk, 119

K

Keyboard operators, 22, 47-48, 138, 143
Keyboards, 37, 41, 107-108
Keybowl, 107-108
Knee jerk test, 160
Kneeling, avoiding, 125-126
Knees, workstation seating and, 95
Knit goods manufacturing, 22

L

Laboratory studies, for low back pain, 161
Layoffs, and CTDs, 109
Layout, of work space, 106-107, 109
Left-handed workers, and risk of CTDs, 49
Letter carriers, 148
Lifting, 104, 122t, 163
 aids to, 46
 forceful manual, OSHA definition of, 44
 heavy, 113t
 reduction of, 131-134, 132f
Lifting index (LI), in NIOSH lifting guides, 82
Lighting, 50, 53, 108
 and computer screens, 103
Litigation, and CTD rate of incidence, 20f
Load constant, in NIOSH lifting guides, 83
Low back pain, 153-167
 bowel/bladder changes in, 164
 chronic
  definition of, 165
  treatment of, 165
 common causes of, 36t, 139t
 cumulative nature of, 111
 foot drop in, 164
 inflammatory vs. noninflammatory causes, 161
 laboratory studies of, 161
 with leg pain, 164
 medication for, 164
 neurological changes in, 161
 physical examinations for, 160
 prevention of, 166
 pulse rate in, 161
 reflexes in, 160
 and renal compromise, 164
 risk factors for, 47
 testing, guidelines for, 162f
 treatment of, 161-165
 and vibration, 49
 work hardening for, 165, 168
Lowering, reduction of, 131-134, 132f
Luggage manufacturing, 148
Lumbar disc disease, 157

Lumbar sacral syndrome, 139t, 153-167
Lumbar support, 126, 133f, 163-164, 167
  belts, 157
  low-cost, 110t
  for lower back pain, 153f

M

Machine pacing, and risk of CTDs, 51
Machine tool operators, 142
MacKenzie exercises, 163
Magnetic resonance imaging (MRI), for low back pain, 146, 161f, 162
Mail workers, 22
Malnutrition, and risk of CTDs, 147f
Manual handling, frequent, OSHA definition of, 44
Manual lifting, forceful, OSHA definition of, 44
Marginal functions
  definition of, 44-45
  samples of, 75f
Massage, for lower back pain, 153f
McMurthey's test, 145
Meat packing workers, 3, 12, 22
Mechanics (profession), 148, 152
Media, and CTD rate of incidence, 20f
Medical conditions, versus behavior problems, 73f
Medication, for lower back pain, 153f, 164
Menses, and risk of CTDs, 143, 147f
Military brace test, 148
Minnesota Workers' Compensation study, 154
Misoprostol, for lower back pain, 164
Monitors, 48
  CTD prevention measures, 134
  placement of, 41
  workstation design and, 100f-102f, 103
Mopping, 113t
Morale, and CTDs, 109
MRI. See Magnetic resonance imaging
Multiple movements, 15, 22
Muscle relaxants, for lower back pain, 164
Muscles, physiology of, 30
Muscle-tendon unit, function and structure of, 30

Muscle wasting, measuring, 158
Musicians, 141

N

Narcotics, for lower back pain, 164
National Institute for Occupational Safety and Health (NIOSH)
  definition of CTDs, 10
  lifting guides, revised, 81-90, 84f-85f
  software vendors for, 84f-85f
Neck
  pain in, 140
  position of, 47
  work station design and, 103
Neck injuries
  common types of, 139t
  vibration and, 49
Nerve compression, avoiding, 125
Nerve conduction studies, for numbness, 31
Nerve entrapment syndromes, upper extremity, 147-151
  risk factors for, 147f
Neurologic changes, in lower back pain, 164
New work, and risk of CTDs, 51, 130
NIOSH. See National Institute for Occupational Safety and Health
Nonsteroidal antiinflammatory drugs, for back pain, 145, 149, 152, 153f, 164
Nortriptyline, for chronic lower back pain, 165

O

Occupational Safety and Health Administration (OSHA). See also Americans with Disabilities Act (ADA)
  definitions
    of CTDs, 10
    of forceful manual lifting, 44
    of frequent manual handling, 44
  enforcement actions, 12
  ergonomics standards, 54, 132f-133f, 134, 185

on signal risk factors, 42-44, 43t
Organization, of work space, 106-107, 109
OSHA. *See* Occupational Safety and Health Administration
Overhead reach, recommended, 17
Overuse injuries. *See* Cumulative trauma disorders
Overwork, and risk of CTDs, 50

P

Pain, measurement of, 31
Painters, 34, 40, 47, 140, 142
Palmaris brevis sign, 151
Patrick's test, 160
Pen/pencil grip, for CTD reduction, 110t
Phalen's test, 32, 36f, 145, 151
Physical Capacities Classification form, 68f-69f
Physical examination
　for lower back pain, 159
　post offer, 62-66, 67f-69f, 70, 73f-74f, 76
　return-to-work, 65
Pinch gripping, and risk of CTDs, 51
Pipe fitters, 43t
Posterior interosseous syndrome, 139t
　diagnosis and treatment, 150-151
Posterior tibial reflexes test, 159
Post-offer exams, 62-66, 67f-69f, 70, 73f-74f, 76
　medical evaluation form for, 67f
Posture, 126, 133f, 140, 148, 154, 163
　assessment of, 52, 52f
　chair adjustment and, 41
　chair design and, 107, 133f
　correct, 41, 85-87, 88f-90f, 92f
　improving, 124
Pottery manufacturers, 22
Power tools, 48
Preemployment testing, 63-65
Pregnancy, and risk of CTDs, 120, 143, 147f
Pre-placement exams, 62-66, 67f-69f, 70, 73f-74f, 76
　medical evaluation form for, 67f

Profession, and risk of CTDs, 22. *See also* specific professions
Pronator syndrome, 139
　diagnosis and treatment, 150
Protruded disk, 112
Ps, seven, of CTDs (mnemonic device), 7
Psychological stress, and CTD risk, 17-18, 20f, 28-29, 154
Pulling, reduction of, 104, 131
Pulse rate, in lower back pain, 160
Pushing, reduction of, 104, 131

Q

Quebec Task Force on Spinal Disorders, 155, 165

R

Radial tunnel syndrome, 139t, 147f
　diagnosis and treatment, 150
Range of motion, 140-142, 149, 158, 166
Rangers, 148
Raynaud's phenomenon of occupational origin, 35
Reaching, reduction of, 113t, 130-131, 132f
Recommended weight limit (RWL), in NIOSH lifting guides, 82
Recovery
　measure of, 154, 156, 164
　timetable for, 154, 157, 158
　in work- vs. non-work-related injuries, 154, 155t
Reflexes, in lower back pain, 160
Renal compromise, in lower back pain patients, 164
Repetitive motion, 15, 47, 51, 122-123. *See also* Cumulative trauma disorders
　definition of, 51
Rest, allowing time for, 123
Return-to-work form, 167f
Return-to-work issues, 138, 155-156, 165-166, 168
Revised layout keyboards, 108

Revised NIOSH Guide Program for Manual Lifting, 84f
　software vendors for, 84f-85f
Rheumatoid arthritis, and risk of CTDs, 147f
Risk of CTDs, 13-15, 28-30, 29f, 64, 147f
　aging and, 19
　alcohol and, 147f, 161
　assessment of, 44-45
　attitude and, 65f
　in back, 47
　cold and, 49
　conflict with supervisors and, 18
　contact stress and, 52, 124-125
　drug use and, 20f, 154, 161
　ethnic groups and, 22
　in extremities, 47, 147f
　fatigue and, 50
　fitness and, 18-19
　industries with greatest, 22
　job rotation and, 119
　lack of control and, 51
　left-handed workers and, 49
　machine pacing and, 51
　menses and, 143, 147f
　new work and, 51, 130
　overwork and, 50
　pinch gripping and, 51
　pregnancy and, 120, 143, 147f
　profession and, 22. *See also* specific professions
　psychological stress and, 17-18, 20f, 28-29, 154
　reduction of, 53-54
　signal risk factors (OSHA), 42-44, 43t
　sleep and, 154
　slippery work surfaces and, 50
　smoking and, 147f, 154
　social factors and, 17-18, 20f, 28-29, 154
　steroids and, 161
　stress and, 17-18, 20f, 28-29, 154
　stress contact and, 52, 124-125
　tilted work surfaces and, 50
　women and, 15-17, 22, 49-50, 119-120, 143, 147f
　work attitudes and, 65f
Riveters, 142
Rotator cuff tendinitis, 34, 36t
　diagnosis and treatment, 142

S

Safety devices, 53
Safety glasses, 53
Salsalate, for lower back pain, 164
Sander operators, 43t
Sciatica, 161, 164
　medication for, 164
Screen, computer. *See* Monitor
Selection programs, to reduce CTDs, 60-62
　legal cautions, 62, 64
Sewer workers, 142
Sewing machine operators, 41
Shipping/loading workers, 43t, 149, 152
Shoe manufacturing workers, 22
Shoulder
　frozen, 36t
　injuries to, common types of, 139t
　workstation design and, 103
Signal risk factors (OSHA), 43t
　definition of, 42-44
Sleep, inadequate, and risk of CTDs, 154
Slippery work surfaces, and risk of CTDs, 50
Slouching, workstation seating and, 95, 96f
Smoking, and risk of CTDs, 147f, 154
Social factors, and CTD risk, 17-18, 20f, 28-29, 154
Social Security Act, and entitlement society, 19
Space, in work site, 106
Spinal manipulation, for lower back pain, 153f, 165
Spinal stenosis, 158, 160, 163
Spinal traction, for lower back pain, 153f
Splints, 145, 152, 168
Split keyboard, 107-108
Spondylitis, 161
Spondylolisthesis
　and lower back pain, 165
　and pre-placement x-rays, 63

Spondylolysis, 163-165
Sports contributing to CTDs, 53
Squatting, avoiding, 125-126
Stage workers, 50
Standing
  alternatives to, 125
  minimizing, 133f
Stenosing tenosynovitis, diagnosis and treatment, 141-142
Steroids
  and CTD risk, 161
  epidural, for lower back pain, 153f, 165
  facet joint injections of, 165
  oral, for lower back pain, 153f
Stockers, 35, 43t, 54, 148-149, 152
Stooping, avoiding, 125-126
Straight leg raise test, 160
Strain index, in NIOSH lifting guides, 84-90
Strength testing, job-specific, efficacy of, 63
Stress
  contact, and CTD risk, 52, 124-125
  and CTD risk, 17-18, 20f, 28-29, 154
Sulindac, for lower back pain, 164
Supervisors, conflict with, and CTD risk, 18
Supraspinatus tendinitis, 34, 36t, 149
  diagnosis and treatment, 142

T

Team approach, to CTD reduction, 4, 80-81, 104, 138. *See also* Workers, cooperation of
Tendinitis
  bicipital, 34-35, 36t, 139t, 147f, 149
  diagnosis and treatment, 141-142
  rotator cuff, 34, 36t, 142
  supraspinatus, 34, 36t, 142, 149
  wrist, 33
Tendons, physiology of, 30
Tennis elbow (lateral epicondylitis), 34, 36f, 147f
  diagnosis and treatment, 141
Tenosynovitis, 36t, 139t
  diagnosis and treatment, 141-142
  first extensor, 152
  stenosing, diagnosis and treatment, 141-142
TENS. *See* Transcutaneous electrical nerve stimulation
Tension myalgia, 139t
  diagnosis and treatment, 139-140
Tension neck syndrome, 36t, 139t
  diagnosis and treatment, 139-140
Thoracic outlet syndrome, 36t, 139t, 147f
  diagnosis and treatment, 148
  risk factors for, 147f
Threat, direct, definition of, 73f
Tile layers, 43t
Tilted work surfaces, and risk of CTDs, 50
Tinel's sign, 32-33, 36f, 145-146, 150-151
Toe guards, 53
Tools. *See also* Equipment
  appropriate, 45, 60
  ergonomic design of, 46
  hand, design of, 126-127
  power, 48
Traction, for lower back pain, 165
Training, of workers, 61-62
Transcutaneous electrical nerve stimulation (TENS), for lower back pain, 153f
Transmission equipment manufacturing, 22
Trauma, and risk of CTDs, 147f
Trazodone, for chronic lower back pain, 165
Twisting, 154, 163, 166
  reduction of, 130, 132f
Typists. *See* Keyboard operators

U

UEMSD (Upper extremity musculoskeletal disorders), 48
Ulnar nerve entrapment, 33, 36f
Ulnar succus syndrome, 151
Ulnar tunnel syndrome, 139t, 146-147, 147f
Ultrasound
  for lower back pain diagnosis, 161f
  for lower back pain treatment, 153f

Uncertainty, diagnostic, 154
Upper extremity musculoskeletal disorders (UEMSD), 48

V

Vacuuming, 113t
Vertical Multiplier (VM), in NIOSH lifting guides, 83
Vibration, 48-49, 122t
 and back and neck disorders, 49
 reduction of, 128-130
 syndromes, 35, 152-153
Vibration-induced white finger (VWF), 35, 48, 152
Vibrator operator, 47
Video display terminals. *See* Monitors

W

Waddell's signs, 160
Walk-throughs, of work site, 118, 138
Warming up, importance of, 62, 65f, 167
Wear and tear injuries. *See* Cumulative trauma disorders
Weight loss, 148, 159t, 161
Welders, 34, 142
Williams' exercises, 163
Window washers, 34-35, 149
Women, CTD risk and, 15-17, 22, 49-50, 119-120, 143, 147f
Word processors. *See* Keyboard operators
Work
 ergonomic analysis of, 46, 104-105
 structure of, 130-131
 unfamiliar, 51, 130
Work areas, ergonomic analysis of, 46, 90f, 103-109

Work attitudes, and risk of CTDs, 65f
Workers
 cooperation of, 4, 65f, 104, 119. *See also* Team approach, to CTD reduction
 ergonomic analysis of, 105
 habit, and CTDs, 20f, 105
 prior CTD history of, 46
 selection programs, to reduce CTDs, 60-62
 legal cautions, 62, 64
 size variance of, designing for, 120
 training of, 61-62
Worker's compensation, 2
Work hardening, for lower back pain, 165, 169
Work site
 design of, 81-82, 118-122
 functional job analysis and, 80-81
 visit to, for CTD risk evaluation, 118, 138
Work stations, 85-87, 88f-90f, 92f
Work surfaces, dimensions and angles of, 121f
Wright's test, 148
Wrist, 143-144, 144t
 position of, 47
 rests for, 41, 45-46, 53, 110t, 125
 tendinitis of, 33
 workstation design and, 95, 98f-100f

X

X-rays
 need for, 157
 pre-placement, efficacy of, 63
 value of, 156-157

Fundamentals of Occupational Safety and Health
By James P. Kohn, et al.

This book covers the basics safety and health professionals need to control hazards, prevent losses, and protect the health and lives of workers. This book balances the management of safety with relevant science and practical aspects of complying with regulations.

Softcover, Index, 429 pages, 1996, ISBN:0-86587-539-1 **$49**

Safety Made Easy: A Checklist Approach to OSHA Compliance
By W.A. Tex Davis, et al.

This book provides a simpler way of understanding your requirements under the complex maze of OSHA's safety and health regulations. Each checklist begins with a brief description of the objectives of the listed items, followed by the required actions and corresponding standards, and where appropriate, training, personal protective equipment, and recordkeeping requirements.

Softcover, Index, 171 pages, 1995, ISBN:0-86587-463-8 **$49**

Total Quality for Safety and Health Professionals
By F. David Pierce, CSP, CIH

Author F. David Pierce explains what Total Quality is and how it can be used to improve the safety process. Using his own experiences and numerous case examples, Pierce examines both the fundamentals and implementation of Total Quality Management. In addition, he discusses common roadblocks to Total Quality and how you can overcome them.

Hardcover, Index, 229 pages, 1995, ISBN: 0-86587-462-X **$59**

"So You're the Safety Director!" An Introduction to Loss Control and Safety Management
By Michael V. Manning, PhD, Manning and Associates

Let author Michael V. Manning's narrative approach and easy-to-follow writing style make it seem like you've hired him to help you start—or upgrade—your safety program, which is exactly what hundreds of companies have done. Manning walks you through the do's and dont's of establishing and evaluating your company's safety program.

Softcover, Index, 174 pages, 1995, ISBN 0-86587-481-6 **$49**

Government Institutes
4 Research Place, Rockville, MD 20850-3226 USA
Tel. (301) 921-2355 • Fax (301) 921-0373
E-mail: giinfo@govinst.com

Ergonomic Problems in the Workplace: A Guide to Effective Management
By James E. Roughton, CSP, CHMM

Now your company can reduce injuries–such as Cumulative Trauma Disorders (CTDs)– and reduce the number of workers' compensation claims. Contents include: developing a program; case histories; hazard assessment; CTDs; workplace hazards; hazard prevention and controls; back injuries and material handling; tool selection; ergonomic personal protective equipment; implementing a program; medical management; VDTs and office ergonomics; training; ADA and ergonomics; and more.

Softcover, Index, 234pages, 1995, ISBN:0-86587-474-3 **$59**

Managing Change for Safety and Health Professionals
By F. David Pierce, CSP, CIH

Pierce has written a how-to book for making change happen in your company's safety and health program. After reviewing Total Quality and its benefits over traditional management, as well as the process of change and reactions to it, he provides a detailed plan for implementing a six-step process for effecting change in your safety program.

Hardcover, Index, 272 pages, May 1997, ISBN:0-86587-563-4 **$59**

Shifting Safety and Health Paradigms
By F. David Pierce, CSP, CIH

In this book, safety visionary and TQM expert F. David Pierce explores how today's safety and health system was developed, what works, what doesn't, and how the practice of safety can and must change in order to better protect workers and contribute to corporate goals into the 21st Century.

Hardcover, Index, 239 pages, 1996, ISBN:0-86587-527-8 **$59**

Understanding Workers' Compensation: A Guide for Safety and Health Professionals
By Kenneth Wolff, D.C. Medical Examiner and Disability Evaluator

Written in layman's terms, this book is designed to help you understand how the Workers' Comp system works, and provide a basic understanding of injury prevention, types of injuries, and cost containment strategies.

Softcover, Index, 180 pages, 1995, ISBN:0-86587-464-6 **$49**

Government Institutes
4 Research Place, Rockville, MD 20850-3226 USA
Tel. (301) 921-2355 • Fax (301) 921-0373
Internet: http://www.govinst.com